Cosmic Questions

OTHER BOOKS BY RICHARD MORRIS

LIGHT

THE END OF THE WORLD

THE FATE OF THE UNIVERSE

EVOLUTION AND HUMAN NATURE

DISMANTLING THE UNIVERSE

TIME'S ARROWS

THE NATURE OF REALITY

THE EDGES OF SCIENCE

COSMIC QUESTIONS

GALACTIC HALOS,
COLD DARK MATTER
AND THE END OF TIME

RICHARD MORRIS

JOHN WILEY & SONS, INC.
NEW YORK CHICHESTER BRISBANE TORONTO SINGAPORE

This text is printed on acid-free paper.

Published by John Wiley & Sons, Inc.

Library of Congress Cataloging-in-Publication Data:

Morris, Richard, 1939–
Cosmic questions : galactic halos, cold dark matter, and the end of time / Richard Morris
p. cm
Includes index.
ISBN 0-471-59521-7
1. Cosmology—Miscellanea. 2. Astronomy—Miscellanea. I. Title.
QB981.M859 1993
523.1—dc20 93-13373

Printed in United States of America

10 9 8 7 6 5 4 3 2 1

CONTENTS

INTRODUCTION

The last ten or fifteen years have seen a revolution in the field of cosmology. But it has been a strange kind of revolution. No widely accepted theories have been overthrown, and few long-standing problems have been solved. On the contrary, there are more unanswered questions in the field today than there were a decade ago.

This does not imply that no progress has been made. One could hardly say that and speak of a "revolution" at the same time. A great deal of new scientific knowledge has been obtained, and some powerful new theories have been developed. At the same time, however, the limits of scientific inquiry have also broadened. Cosmologists and astrophysicists have begun to ask questions that would not even have been considered "scientific" a few years ago. As knowledge has increased, the mysteries have deepened, and scientists have begun to realize that our universe is a more baffling thing than they had dreamed it could be.

As the decade of the 1970s came to an end, cosmology appeared to be a relatively straightforward subject. Some very successful theories about the structure and evolution of the universe had been developed. The scientific data that had been gathered indicated that the universe had its origin in an enormous explosion called the "big bang," which took place 10 or 15 billion years ago. But what caused the big bang? In the view of many scientists, this was a question that might remain forever beyond the grasp of physics.

However, it did seem possible to speak with confidence of events that had taken place when the universe was only a few seconds old. The events that had taken place during that epoch had left a permanent imprint on the cosmos, and their character could be deduced.

Furthermore, the evolution of the universe since that time could apparently be described by Einstein's law of gravitation, called the "general theory of relativity." During the 1960s and 1970s, numerous experiments had been performed which confirmed Einstein's theory to a degree of accuracy that would have been inconceivable a few years earlier. Thus scientists were confident that they could use general relativity to describe the structure and evolution of the universe and to speculate about the path along which it would evolve in the future.

But then, as scientific knowledge increased, it became clear that matters were not so simple. Astronomers found that the universe contained enormous quantities of a kind of matter that they could not see. As more observations were made, it gradually became apparent that at least 90 percent of the mass of the universe consisted of this invisible "dark matter." But no one knew what dark matter was. In fact, there seemed to be good reasons for believing that this mysterious substance was something that was entirely unlike ordinary matter.

In 1980, physicist Alan H. Guth proposed his inflationary universe theory, and scientific knowledge took another leap forward. Guth's new conception, which was an elaboration of the big bang theory, allowed scientists to speak of events that had taken place when the universe was only a tiny fraction of a second old, and to understand certain puzzling features of the present-day universe as well.

Every time a question was answered, however, several new ones were sure to appear in its place. The inflationary universe theory allowed scientists to look back to a time so close to the beginning that some of them began to wonder if it might not be possible to go a step further and to speculate about the events that caused the universe to come into being. Meanwhile, other scientists went even further and used the inflationary universe theory as a springboard for speculation about the possible existence of other universes.

At the same time, the theory made the dark matter problem appear even more puzzling, for it seemed to imply that the dark matter could not be composed of protons, neutrons, and electrons, the particles that made up the objects encountered in our everyday world and that are found in such astronomical objects as stars and plants as well.

As knowledge expanded, so did the number of unanswered questions. Scientists found that there was more to this thing called "physical reality" than they had thought, and they found themselves posing queries which had previously been considered questions for metaphysics or theology, or perhaps not even meaningful questions at all.

In his 1977 book about the origin of the universe, *The First Three Minutes,* Nobel prize–winning physicist Steven Weinberg observed that "in the 1950s, the study of the early universe was widely regarded as the sort of thing to which no respectable scientist would devote his time." Weinberg then went on to point out that by the time he began writing his book, matters had changed and the history of the early universe had become a topic of intense scientific discussion.

Since Weinberg wrote those words, matters have changed even more dramatically. Nowadays scientists not only discuss events that took place in the early universe, but they ask what, if anything, might have been happening before there was a universe. They ask whether there might not be other universes that are connected to our own through microscopic "wormholes," and whether it might not be possible for universes to reproduce themselves. They pose queries about the origin and the nature of time and ask whether it might not be possible under certain circumstances to travel into the past.

And, like all scientists, they look for connections. When they do, they often find mysterious links between microcosmos and macrocosmos, between discoveries in the field of high-energy particle physics and our understanding of the cosmos as a whole. And they often find the reverse: Discoveries in cosmology sometimes turn out to have implications for scientists' understanding of the behavior of elementary particles.

Meanwhile, scientists working in a new field called "quantum cosmology" are seeking to find connections between the manner in which the universe came into being and the properties that it exhibits today. Some of them have pointed out that, in many ways, our universe is very special, and that if it had not had "initial conditions" of a particular character, it would never have been capable of supporting life.

In this book, I will describe some of the "cosmic questions" that are found at the heart of contemporary cosmological speculation. I will discuss the attempts that have been made to solve the cosmic mysteries that have appeared during the last decade, and I will discuss some attempts to go even further and to speculate about the nature of creation itself.

It may be that I pose more questions than answers in this book. If I do, I make no apologies. Questioning, after all, is the essence of science. It is no accident that the best scientific minds have always tended to be attracted to those fields in which the mysteries seemed most baffling. Some of those that are encountered in the field of cosmology are baffling indeed. And it is this fact which makes cosmology the most exciting field in contemporary science.

1

Did the Big Bang Really Happen?

On April 23, 1992, at a meeting of the American Physical Society in Washington, D.C., a team of scientists headed by George Smoot of the University of California at Berkeley announced a finding that future generations will probably consider one of the important scientific discoveries of the twentieth century. Using instruments that had been placed on NASA's Cosmic Background Explorer (COBE) satellite, they had been able to look far back in time and had succeeded in measuring the temperature of the universe when it was only about 300,000 years old.

At that time—10 to 20 billion years ago—the universe was still in its infancy. Only a short time before (at least a short time by astronomical standards), it had been born as an inconceivably dense, hot fireball. Although it had cooled considerably during its first 300,000 years, it was still very hot by terrestrial standards. The glowing hydrogen and helium gas which filled the young universe had a temperature that was about equal to temperatures found on the surfaces of red giant stars. But the 300,000-year-old universe contained no stars. Another billion years or so were to pass before the hot gas would begin to condense into stars and galaxies.

Scientists had been trying to find evidence of density fluctuations in the early universe for years. They knew that the galaxies and clusters of galaxies that were observed by astronomers could only have been created if the universe had been denser in some places than in others. There simply wasn't any other way it could have happened. Regions of higher-than-normal density would have exerted a gravitational attraction on other matter in their vicinity, pulling the surrounding matter in, and would thus have become denser yet. Without such gravitational "seeds" the universe would never have contained anything but diffuse gas. If there hadn't been high-density regions when the universe was very young, then the present-day universe would have no stars, no galaxies, and almost certainly no life.

Smoot and his colleagues had been able to determine that when the universe was 300,000 years old, certain small regions had been one-thousandth of 1 percent hotter than other regions. And the only way to explain these temperature "bumps" or "ripples" was to conclude that the early universe had been denser in some places and more rarefied in others.

The existence of such small temperature variations may not be impressive to the reader who is unfamiliar with such matters. However, such small density fluctuations that would grow in time were really all that was needed. In fact, if the scientists who had been studying the early universe had discovered variations that were much larger, that would have been troubling. Fluctuations that had originally been too big would have led to structures in the present-day universe much larger than those that were actually observed. Calculations indicated that fluctuations of the size observed would produce galaxies of the size observed today.

Not only could the temperature variations be linked to the subsequent evolution of the universe, but their discovery also provided clues about the nature of the universe when it was much younger than age 300,000. The findings seemed to provide support for a description of the early universe known as the inflationary universe theory.* They provided scientists with their first peek at a very early "quantum era" as well. There was good

---

* This theory will be discussed in Chapter 4.

reason to believe that the fluctuations seen by Smoot and his colleagues had originated as submicroscopic events taking place when the universe was only a small fraction of a second old.*

PROOF FOR THE BIG BANG?

The discovery made by Smoot and his colleagues was widely heralded as a "proof" of the big bang theory of the universe. In a way it was. At last, scientists had observed the universe when it was very young and expanding rapidly. However, the big bang theory really didn't need to be bolstered in this manner. Though there had been some premature (and sometimes misleading) reports of the theory's demise, it was still on very firm ground when the COBE findings were reported. To be sure, it had encountered some difficult problems. However, these had centered around such phenomena as galaxy formation. In the view of most scientists, there were no good reasons to question the basic theory. Only a few dissenting scientists had embraced the position that it was necessary to look for alternatives to the idea that the universe had begun as a dense, hot fireball billions of years ago.

Scientists, of course, have a habit of questioning everything. Because they know that accepted scientific ideas have been overthrown many times in the past, they take nothing for granted, and they look for "holes" in even the most convincing arguments. It is a measure of the success of the big bang theory that it emerged unscathed from such questioning. Even if the COBE team had failed to find the fluctuations they were seeking, it is doubtful that the theory would have been overthrown and doubtful that it will be in our time; the support for it is simply too solid.

There are at least three different kinds of scientific evidence that would make any alternative to the big bang idea seem very unlikely. I will consider them one by one.

The Cosmic Microwave Background Radiation

In 1964, while working with a radio antenna at Bell Laboratories in New Jersey, two scientists, German-American physicist Arno

---

* And I'll have a lot to say about the very early universe in subsequent chapters too.

Penzias and American radio astronomer Robert Wilson, discovered that a strange kind of radiation was falling on the earth. This radiation was coming from every direction in the sky, and it never varied in intensity; it was equally strong at every hour of the day or night. Nor did it vary according to direction. The radiation that was coming from a region near the Big Dipper was no stronger or weaker than that which seemed to originate in regions near the star Sirius, for example. And, as measurements that would subsequently be made by other scientists were to show, the radiation was just as strong in the earth's southern hemisphere as it was north of the equator.

The two scientists were not particularly alarmed by what they had discovered. There was nothing dangerous about the mysterious radiation. In fact, it consisted of nothing more than a background of very-low-intensity radio waves.* In fact, the intensity of the radiation was so low that they weren't even surprised that no one had noticed it before.

However, Penzias and Wilson were intrigued. The existence of the radiation was a puzzle that had to be explained, and it wasn't long before an explanation was forthcoming. Penzias and Wilson soon discovered that the existence of just such background radiation had been predicted by the Russian-American physicist George Gamow and his collaborators Ralph Alpher and Robert Herman some fifteen years previously. In 1949, Gamow, Alpher, and Herman had pointed out that, if the universe had begun with a big bang, then the glow of the big bang fireball should still be visible today. It should exhibit the form of blackbody radiation at a temperature of about 5 K, and it should fall on the earth from every direction in space.

Before I explain the relationship between blackbody radiation and the big bang, it will be necessary to explain a few terms. The symbol *K* stands for *kelvins,* and 5 K is five degrees above absolute zero on the Celsius scale. Absolute zero is the lowest possible temperature. It is the temperature at which all molecular motion ceases, and it is equal to –273°C, or –460°F. 5 K is thus equal to –268°C. Though the kelvin and Celsius scales are equivalent in the sense that they use degrees of the same size, it is often more convenient to use the former because one can then avoid having to use negative numbers.

---

* Naturally, these were not a radio broadcast. There are many natural processes that create radiation in the radio frequency band.

Blackbody radiation is radiation that would be emitted by an object that was perfectly black. Though perfect black bodies do not exist (every object will reflect at least some small percentage of the light that falls on it), it is possible to construct apparatus that behaves like a blackbody in the laboratory, and the characteristics of blackbody radiation have been studied experimentally as well as theoretically.

All objects emit radiation of some kind. For example, an iron rod will exhibit a red glow if it is heated, and the glow will turn white if it is heated still more. Objects at room temperature do not emit visible light, but they do give off infrared radiation which can be measured. Very cold objects radiate too. One that has a temperature of a few degrees above absolute zero will give off short radio waves called microwaves.

What Penzias and Wilson had seen was the glow of the big bang fireball. As it turned out, the temperature of this radiation wasn't exactly what Gamow and his collaborators had predicted. It is 2.7 K rather than 5 K. However, the very fact that blackbody radiation coming from everywhere could be seen was a confirmation of the hypothesis that the universe had begun with a big bang.

At this point, one might ask why it was radio waves that Penzias and Wilson observed when the universe had originally given off a brilliant glow of light. As it turns out, there is more than one way of looking at this question. Consequently, I will begin by making an analogy; then I will rephrase the answer in more precise terms.

The reason that Penzias and Wilson saw radio waves was that billions of years had passed since the big bang, and the universe had cooled off considerably during that time. What had originally been a brightly glowing object (yes, it is perfectly correct to refer to the universe as an "object") had dimmed like a dying ember. As the eons had passed, its glow had become dimmer and dimmer until there was nothing left but microwave radiation.

Another way of saying this would be to point out that the universe had been expanding (a discussion of the expansion of the universe will come shortly) for billions of years. As it expanded, the light that traveled through it expanded too; wavelengths became longer. The first thing that happened was that the visible light was transformed into infrared radiation; rays in the infrared band have longer wavelengths than those that make up the visible spectrum. Then, after more time passed, the infrared radiation changed into radio waves, which were longer yet. The microwaves that were seen in the sky were light waves that had been greatly stretched.

But why did the microwaves come from everywhere, and not from some particular location in the sky? The reason was that the big bang had been something that happened everywhere. When the universe was born, it did not explode outward into some preexisting empty space. The big bang filled the entire universe because it *was* the universe. As the universe expanded, so did the volume of space. The microwaves came from everywhere because the region of space in which the earth was located was originally inside the big bang.

Since Penzias and Wilson made their discovery, the cosmic microwave background has been studied extensively. It has been observed with radio telescopes on the surface of the earth, with instruments sent aloft in rockets, and with apparatus placed on satellites. These observations have established that the microwave background varies in intensity with wavelength in exactly the way that blackbody radiation should. The fact that it has the characteristic "signature" of blackbody radiation makes it seem extremely unlikely that it could have been produced by anything but the big bang. At least, no other plausible explanations for its existence have ever been suggested.

The Chemical Composition of the Universe

The universe is made up primarily of hydrogen and helium. All of the other elements, including such important ones (important to human beings, anyway) as carbon, nitrogen, oxygen, and iron, exist only in relatively tiny quantities. In the cosmic scheme of things, they are nothing but trace impurities.

The universe is approximately 25 percent helium and 75 percent hydrogen. There are about ten hydrogen atoms for every atom of helium, but a helium atom is about four times as heavy as one of hydrogen, so a little helium goes a long way,

These percentages do not vary much from place to place. Stars are about 25 percent helium, and so are the clouds of gas that can be seen in interstellar space. Astronomers observe roughly this quantity of helium in old stars, in relatively young stars, and in those brightly glowing objects known as quasars. Even though quasars tend to be billions of light years away, it isn't terribly difficult to study their chemical composition. Scientists can do this by analyzing the light that they emit.

Helium is also found in cosmic rays. Cosmic "rays" are not really radiation. On the contrary, they consist of rapidly moving particles which are thought to be present throughout the universe. Some of these particles are hydrogen nuclei, or protons; others are helium nuclei, also called alpha particles. The ratio of hydrogen to helium in cosmic rays is about the same as it is everywhere else.

At first, one wouldn't think that it would be difficult to explain why there should be so much helium in the universe. Explaining the existence of hydrogen is no problem. Hydrogen atoms consist of nothing but protons and electrons and will form spontaneously. Helium, on the other hand, has a more complex character, and must be created in nuclear reactions. Stars are powered by nuclear fusion reactions, and these reactions, which are similar to those that take place in a hydrogen bomb explosion, convert hydrogen into helium. These multistep reactions bring together four protons, or hydrogen nuclei, to create a helium nucleus.

Scientists are confident that they understand these fusion reactions in detail, and they can calculate how much hydrogen should have been converted into helium during the billions of years that have passed since the universe began. But when this calculation is performed, one does not obtain a figure of 25 percent. It turns out that the amount of helium that could have been introduced in stars is no more than a few percent.

The only conclusion that one can draw is that if the universe is 25 percent helium now, then it must have contained nearly 25 percent helium a short time after its beginning. And this helium must have been created in some kind of reaction. The assumption that the universe might have begun as 25 percent helium just doesn't work. When the universe was less than one minute old, helium could not have existed. Calculations indicate that, before that time, temperatures were too high and particles had too much energy. If any helium nuclei were created then, they must have immediately been blasted apart.

On the other hand, a great deal of helium could easily have been synthesized after the universe was about one minute old. By this time, temperatures had decreased to a point where the protons and neutrons of which a helium nucleus is composed would hold together. There would still have been collisions with other particles, but they would not have been so disruptive.

And then, as the universe continued to expand and temperatures continued to drop, the reactions that produced helium gradually came to a halt. The average energy of each nuclear particle was smaller, and matter was more dispersed. Under these conditions the process of helium production slowed and finally ceased.

If one makes the assumption that there was a big bang, then it is easy to construct scenarios for the creation of helium. But this does not really constitute proof that a big bang took place. At best, one has only shown that the hypothesis of a big bang is not contradicted by observations of the chemical composition of the universe.

However, there is a related body of evidence that does seem to confirm a big bang origin for the universe. This has to do with the fact that the universe contains small quantities of a substance called deuterium. Deuterium is an isotope, or variety, of hydrogen. Most hydrogen nuclei consist of single protons. A hydrogen atom is normally composed of a proton with a single electron in orbit around it. The positive charge on the proton and the negative charge of the electron balance one another in the electrically neutral hydrogen atom.

A deuterium nucleus, on the other hand, is made up of a proton and a neutron that are bound together. Deuterium is an isotope of hydrogen because the addition of an electrically neutral neutron does not alter any of the atom's chemical properties. There is still only one electron in orbit, and in a chemical reaction, deuterium and ordinary hydrogen will behave similarly. The only significant difference is that an atom of deuterium is heavier.

Observations indicate that deuterium is present in our universe in a concentration of about fifteen parts per million by weight. This doesn't sound like much. However, the very fact that any deuterium exists at all has great significance.

Unlike helium, deuterium cannot be made in stars. The proton and neutron that make up the deuterium nucleus are weakly bound to one another. The high temperatures that exist in stellar interiors would cause such a nucleus to break apart almost as soon as it was formed. The first energetic particle that happened by would knock the neutron and the proton apart. Therefore, deuterium must have been created in some other way. The only plausible idea that has been suggested is that it was made in the big bang, at about the same time that helium was manufactured.

Under such conditions, particle collisions were frequent enough to make deuterium, but temperatures had fallen to low enough a level that it could survive.

There are other light elements that exist in small quantities, and similar arguments can be made about them. Like deuterium, they could only have been made in the big bang. One of these is helium 3 (here the numeral *3* indicates that the nucleus of this isotope contains three particles: Two protons and a neutron). Another is lithium 7, an isotope of light metal lithium.

For the sake of completeness, I should probably comment on the formation of the heavier elements, including such important constituents of our bodies as carbon, oxygen, and nitrogen. The existence of these substances has little bearing on the big bang theory. Astronomers believe that all of the heavier elements were created in nuclear reactions that took place in stars, and that they were subsequently spread through space when some of these stars were destroyed in catastrophic events called supernova explosions that blow a dying star apart. According to current scientific thinking, our planet and almost everything on it, including human beings, is constructed from cosmic debris.

The Expanding Universe

The very first item of evidence obtained by scientists which indicated that the universe began with a big bang came long before the measurements of the concentrations of light elements in the universe, and also long before the discovery of the cosmic microwave background. This was the discovery, made by the American astronomer Edwin Hubble in 1929, that the universe was expanding.

I've saved the discussion of Hubble's discovery for last, not because it's more complicated than the phenomena previously discussed, but simply because I wanted to discuss the evidence for the big bang origin of the universe before beginning a discussion of the universe's present-day structure. As will soon become apparent, observations of the expansion of the universe not only indicate that there was a big bang long ago, but also have a bearing on such matters as the age, density, and composition of the universe, and on the question of what its ultimate fate will be.

It is probably correct to say that Hubble was the scientist who discovered the big bang; at least, he was the individual who first suggested that the universe had a very definite age, that it had not always existed. In 1929, even such scientists as the great Albert Einstein tended to think that the universe was static and had probably existed forever. This view had prevailed since the time of Aristotle, and it had never been seriously questioned, at least not within a scientific context. To be sure, scientists had found that Einstein's theories seemed to imply that the universe might have begun at a particular point in time and had been growing ever since. However, even Einstein refused to take this possibility seriously.

After Hubble announced his discovery, however, Einstein quickly realized that the traditional conception of the universe was untenable. In fact, he went so far as to describe his previous insistence on a static universe as "the greatest blunder of my life." For he immediately understood the significance of Hubble's discovery. The universe hasn't remained the same. It has evolved.

What Hubble had observed was that, with a few exceptions, all of the galaxies were rushing away from the earth. Furthermore, the greater the distance between the earth and a galaxy, the faster the galaxy was receding. These astronomical observations did *not* indicate, however, that the earth or its solar system was the center of the universe. The reason that the galaxies seemed to be traveling away from the earth was that they were all moving away from one another. The phenomenon that Hubble had observed was one that would be seen by any astronomer in any galaxy of the universe.

A number of different analogies have been invented to illustrate this point. My favorite is a rising loaf of raisin bread. Imagine that a lump of raisin bread dough is placed in an oven. As the bread rises, the dough expands, and all the raisins recede from one another. If two raisins are initially very close, the distance between them will not change very rapidly. If they nearly touch one another at first, then they will still be fairly close to one another when the bread is taken out of the oven. But if two raisins are initially farther apart, the size of recession will be greater. For example, the distance between two raisins on opposite sides of the loaf will change noticeably.

Every analogy breaks down at some point, and this one is no exception. As we shall see in the next chapter, the universe possesses no boundaries that would correspond to the edges of a loaf of bread. No one knows whether the universe is infinite or finite. Indeed, this is a question that may never be answered. But scientists do know that in neither case would there be any place where the universe "ends." An infinite universe would have no edges for the simple reason that space would go on forever. And if Einstein's theory of gravitation (this theory, called the general theory of relativity, will be discussed in more detail later) is correct, then a finite universe would curve back upon itself in a manner similar to the way in which the surface of the earth is curved. No matter how far a ship sails, it never comes to the end of the world. A space vehicle could never come to the edge of the universe either. There simply isn't any such thing.

Some of the galaxies that astronomers observe today are flying away from the earth at velocities approaching the speed of light. However, even when they move that fast, their motion cannot be observed directly because the universe is too big. Intergalactic distances are so vast that the images of even the most rapidly moving galaxies do not change in the period of a human lifetime. If an astronomical plate shows a galaxy in a certain position now, an astronomical observation made a century from now will still show it in the same place.

Fortunately, there is another way to measure the velocity at which a light-emitting astronomical object is approaching or receding. This method, which is quite straightforward, can be applied to any object that emits radiation. One can use it to measure the velocity of a star, a galaxy, or a cloud of interstellar gas.

All forms of radiation, including light, radio waves, infrared and ultraviolet radiation, x-rays, and gamma rays, consist of oscillating electromagnetic fields. Like ocean waves, these electromagnetic waves have crests and troughs. When an object is approaching the earth (or when the earth is moving toward it), the wave crests and troughs will "bunch up," because each time a wave crest is emitted, the object will be a little closer to the earth. If it is visible light that is being observed, the wavelengths will be shifted toward the blue end of the visible spectrum and the light will be said to be blue-shifted, because the blue and violet wavelengths of visible light are shorter than those in the rest of the spectrum. The definition of *wavelength,* by the way, is just what one would expect it to be: The distance between two successive wave crests.

If, on the other hand, an object is moving away from us (or if we are moving away from it), the light waves will become "stretched out"; successive wave crests will be farther from one another. It is easy to see why. If an object (star, galaxy, flashlight, whatever) is at a certain distance when a first wave crest begins to make its way toward us, then it will be farther away when the new one is emitted. The second wave crest will have a longer distance to travel and the distance between two successive crests will be greater. This can be made clear by means of the following analogy. Suppose that a beeper emits sounds at 1-second intervals, and that it is placed 10 feet away. One will obviously hear beeps that are 1 second apart. Now suppose that the beeper is moving away from a listener at a rapid velocity. I will assume that when it emits the first beep, it is 10 feet away, when it emits a second beep, it will be 120 feet away, and so on. The beeps will now be heard at intervals of about 1.1 seconds because it will take the sound an extra tenth of a second to travel the extra 110 feet.

Since the longest wavelengths of visible light are those associated with the red end of the spectrum, light from receding objects is said to be redshifted. The light from most observed galaxies exhibits such a shift. And the greater the redshift, the faster the galaxy can be assumed to be moving away.

One should not conclude, by the way, that light from distant galaxies therefore looks red. It doesn't because visible light is not the only kind of radiation galaxies emit. While yellow or green wavelengths may be transformed into red ones, wavelengths of invisible ultraviolet light will simultaneously shift into the visible part of the spectrum. Thus, to the eye, light from a rapidly speeding galaxy will look pretty much the same as light from one that is hardly moving at all.

However, when the light is examined with scientific instruments, its appearance is quite different. Every chemical element emits light of certain very specific wavelengths when it is heated. Since the light that comes to us from distant galaxies has its origin in such hot objects as stars, it is possible to learn a great deal by breaking it down into its component wavelengths. One can determine not only the velocity of recession of a distant object but also its chemical composition. It is possible to tell how much helium is present, for example, by looking at the (redshifted) wavelengths that show helium's characteristic "signature."

The only galaxies that sometimes exhibit blueshifts are certain members of a small cluster of about twenty galaxies known as the "local group." These galaxies, which include our own Milky Way galaxy, are gravitationally bound to one another. They revolve around one another much like the planets revolve around the sun. Their motion is more complicated—there are more of them than there are planets, and none of them outweighs the others to the extent that the sun dominates the planets—but the basic principles are the same. The galaxies in the local group stay together and thus do not participate in the general expansion of the universe. This is quite a common situation. The universe seems to be full of clusters of galaxies of varying sizes.

The largest galaxy in the local group is the great galaxy in Andromeda. Photographs of it have appeared on numerous posters and have been printed in hundreds, possibly thousands, of books. At the moment, the Andromeda galaxy is moving toward the Milky Way, and so its light is blueshifted. We needn't fear, however, that the two galaxies will collide (even if this did happen, it would not be a particularly catastrophic event; galaxies, after all, are mostly empty spaces). Andromeda and the Milky Way are simply orbiting around one another. There will be other times when they are moving apart.

If all that Hubble had observed was that most galaxies were moving away from the earth, he would not have learned very much. Scientists generally try to discover numerical relationships between quantities they can measure. Only if they succeed in doing this can they make the quantitative predictions that allow a theory to be tested. For example, one could not test Newton's law of gravity simply by observing the fact that the moon revolves around the earth. It is necessary to obtain definite numbers—the distance to the moon and its velocity would be sufficient—that can be "plugged into" the theory. Only when this is done and all the relevant calculations made, is it possible to tell whether a theory really works. Similarly, one could not relate the gravitational attraction of the earth and the moon to the effect of gravity on a falling apple if one did not measure the distance through which the apple traveled and the time it took to make its way to the ground from a branch on the tree.

So Hubble needed to find a mathematical relationship between the recession velocities of galaxies other than the local group and some other quantity. The obvious one was distance. Only after Hubble had measured distances could he show that the farther away a galaxy was, the faster it would seem to be moving.

And only after he had shown that could he conclude that the universe was in a state of expansion.

Intergalactic distances are hard to measure, however. There are so many uncertainties involved that distance measurements are still a source of scientific controversy today. Furthermore, these distance uncertainties have created corresponding uncertainty about the true age of the universe. No one knows precisely how far away the more distant galaxies are. As a result, it is not possible to calculate exactly how fast the universe is expanding, or when the expansion began. That is why I have given the rather imprecise estimate of "10 to 20 billion years" for the age of the universe. All that one can say for certain is that its true age is probably somewhere in this range.

The uncertainty about the age of the universe is a topic that I will return to from time to time. At that time, I will describe the difficulties associated with measuring intergalactic distances in greater detail. For now, I will confine myself to pointing out that, even though there is much about which scientists remain uncertain, they are convinced that Hubble was right when he maintained that the universe was expanding.

It is true that it is often impossible to determine with absolute precision exactly how far away a pair of galaxies—I will call them galaxy A and galaxy B—may be. However, *relative* distances can usually be measured with fair accuracy. One might determine that galaxy B is 2.7 times as far away as galaxy A, for example. Thus it has been possible to establish that there does exist a relationship between the distances of galaxies and their velocity of recession. Galaxies that are twice as far from the earth as certain others move away at speeds that are twice as great. Galaxies that are ten times the distance of some arbitrary reference galaxy have velocities of recession that are ten times as great. This relationship between distance and velocity of recession is known as "Hubble's law." Astronomers observe deviations from this law only in the case of very distant galaxies, which move at velocities approaching the speed of light.

What Hubble's law tells us is that the universe is expanding, and furthermore, that the expansion is uniform. And if the galaxies are flying away from one another in a uniform manner now, then it seems natural to conclude that there was a time when all the matter in the universe was confined in a very small space. If one takes the present expansion of the universe and extrapolates backward, it is difficult to avoid the conclusion that the universe had a beginning. And when one calculates when the universe did begin—taking the various distance uncertainties into account—one can conclude that this beginning lies 10 to 20 billion years in the past.

DID THE COBE EXPERIMENT PROVE THAT THERE WAS A BIG BANG?

No, it didn't, because the big bang theory didn't need any additional support. When George Smoot and his colleagues reported their results, there were already three convincing kinds of evidence that a big bang had taken place: The existence of the cosmic microwave background; measurements of the abundances of such light elements as helium, deuterium, and lithium; and, finally, observations of the expansion of the universe. It is practically inconceivable that a universe which exhibits such characteristics could have begun any other way than in a big bang.

When I say this I am not downplaying the importance of the COBE measurements of the fluctuation in the microwave background. In fact, I quite agree with those scientists who believe that these results will usher in a "golden age of cosmology." For the first time, it has been possible to "take a picture" of the universe as it looked only a few hundred thousand years after it was created, and the picture has confirmed that, even at a relatively young age, the universe contained the kind of structure that astronomers had expected it to have.

But if we are indeed on the verge of a golden age, it is not because the COBE experiment has answered a lot of questions. As we shall see, the problems that have been cleared up are relatively few. If anything, the COBE experiment has increased the number of unsolved problems.

But this is precisely why the experiment was important. Now that scientists have a way to directly observe the early universe, we can expect that new lines of scientific inquiry will open up. And, after all, inquiry is what science is all about. Science flourishes when there are baffling new problems to be solved, and it stagnates when the problems are relative few. Questioning, after all, is the essence of science.

I hope I'm not becoming preachy when I say this. I find that many lay people seem to think of science as the accumulation of knowledge that often hardens into dogma. Indeed, it does sometimes exhibit this characteristic. However, the ages of great scientific discovery are always the ones in which scientists encounter problems about which they understand little or nothing.

As we shall see, the COBE results and other recent astronomical discoveries have not allowed scientists to tie up cosmological knowledge in any neat little packages. In fact, they still are not sure how old or how big the universe is. No one can yet say precisely what its ultimate fate will be. Scientists aren't even sure what 90 or 99 percent of the matter that exists in the universe is made of.

They still haven't figured out how galaxies were created, or how certain large structures—great clusters of galaxies, for example—came about. They are sure that there was a big bang; at least there are no reasonable alternatives to that idea at present. But they don't yet know why the big bang happened, or whether there was anything before the big bang. Some of them suspect that our universe may not be the only one that exists. Indeed, it has even been suggested that there might be an infinite number of universes. But of course no one knows the answers to questions associated with that idea either.

It seems that the field of cosmology is full of questions, and that there may be more of them than there are definite answers. That is precisely what impelled me to write this book.

2

Is Space Infinite?

Does the universe go on forever, or does space have a finite volume? In principle, this question could be answered. Indeed, numerous attempts have been made; scientists have been trying to solve the problem of the "size" of the universe at least since the time of Isaac Newton. On a number of different occasions, they have thought that they have found the answer. However, in every case, new discoveries were made that cast doubt on the conclusions that scientists had reached.

One thing that is clear is that the observable universe is only a small part of the whole. If the universe is 15 billion years old, then it is impossible for astronomers to see objects that are more than 15 billion *light years* away. This follows from the definition of the term light year, the distance that a ray of light will travel in one year. If the universe is 15 billion years old, then the light from objects that are 25 billion light years away won't reach us for another 10 billion years. Obviously there is little hope that human beings will ever observe them. The earth won't even exist 10 billion years from now; the best estimates indicate that we only have about 5 billion years before the sun expands into a red giant and turns our planet into vapor.

There is much more to the universe than we can see. If scientists are to determine how much more there is, they must therefore use indirect methods. Naturally there is nothing wrong with doing this. After all, the existence of many of the phenomena with which physics and many other sciences deal is established in such a manner. No one has ever seen an electron, for example. It is far too small an object to be observed with the eye, or even with the most powerful microscopes. But no scientist doubts that the electron exists. It plays too important a role in far too many observable phenomena.

Before I begin to describe the attempts that have been made to determine whether the universe is infinite or finite, I should clarify one point. An infinite universe can begin with a big bang. All that the big bang theory tells us is that the matter in the universe was initially very compressed. When we say that the universe is expanding, all that we mean is that the matter contained in it is becoming progressively more dispersed. The big bang, one should recall, was something that happened everywhere.

NEWTON'S ARGUMENT

One of the first scientific attempts to determine whether the universe was infinite or finite was made by Sir Isaac Newton in 1692. The universe was obviously infinite, Newton argued in a letter to the English clergyman Richard Bentley. If it were finite, he explained, then gravity would cause all of the matter in the universe to collect at its center. One would not see a large number of individual stars; there would be nothing but a single huge mass. On the other hand, Newton went on, matter would be more or less evenly distributed in an infinite universe, because gravitational forces would not tend to pull bodies in any particular directions.

Unfortunately, Newton's argument was wrong. If it were correct, a phenomenon of this sort could be observed in galaxies (which are finite). All the matter that a galaxy contained—stars, gas, interstellar dust—would fall into the galactic core.

Obviously, this does not happen. If it did, we wouldn't even be here to argue about the matter. All that it takes to keep a galaxy from collapsing is a little rotation. The stars in a galaxy revolve around the galactic core much like the planets of our solar system revolve around the sun. There is no more chance that they will fall into the center of the galaxy than there is that the planet Mars will suddenly leave its orbit and drop into the sun.

Newton was also wrong in assuming that matter in an infinite universe had to be evenly distributed. This is definitely not the case. Once there was a small excess of matter in one region or another, gravitational forces would draw yet more matter into that region. In fact, as I have pointed out previously, this is precisely how galaxies must have been created.

The flaw in Newton's argument was the assumption that if forces were evenly balanced initially, they would remain evenly balanced. This is not necessarily the case. If one were to balance a pencil on end, for example, there might initially be no forces that would cause it to fall one way or another. But it does not follow that the pencil must therefore remain balanced for all eternity. It won't. A very small disturbance, such as a gust of air, is sufficient to send it toppling in one direction or another. Similarly, if one had an infinite universe that had stars that were roughly equidistant from one another, they would not remain in such a configuration. The smallest disturbance in such an arrangement would cause the stars to begin to collect into groups.

Twenty-eight years after Newton's conjectures, in 1720, another attempt was made to reach a conclusion about the "size" of the universe. This time, however, the opposite conclusion was reached. According to Edmund Halley, the English astronomer after whom Halley's comet was named, the universe was obviously finite. "If the number of Fixt Stars were more than finite," Halley said, "the whole superficies of their apparent Sphere would be luminous." Halley had concluded that an infinite universe would contain an infinite number of stars, and that these would give off so much light that the night sky would be very bright. In fact, if Halley's argument were correct, the sky would be as bright as the sun itself. After all, in whatever direction one looked, one's line of sight would inevitably converge on a star.

Today, Halley's argument is known as "Olbers' paradox," after the German astronomer Heinrich Olbers, who resurrected it in 1826. Obviously, Olbers doesn't deserve credit for the idea, for the paradox was stated, not only by Halley, but also by the Swiss astronomer Jean Philippe de Cheseaux, long before Olbers was born. However, "Olbers' paradox" is nevertheless what it is called.

Unfortunately, this argument really doesn't allow us to draw any conclusions about the size of the universe either. Arguments based on Olbers' paradox don't hold up because our universe is not in fact infinitely old. If we assume that the big bang took place 15 billion years ago—15 billion is often used because it's a compromise between 10 and 20 billion—then there are only a finite number of visible stars in an infinite universe. We simply cannot see stars that are more than 15 billion light years away. Olbers' paradox begins with the assumption that, in an infinite universe, the light from an infinite number of stars would fall on the earth.

EINSTEIN'S THEORY OF GRAVITATION

Attempts to determine the "size" of the universe using arguments based on general principles had failed. Eventually scientists realized that they needed more information before they could entertain any hopes of deciding whether the universe was infinite or finite. A theoretical advance of some kind must be made, without which astronomers did not know what to look for. They did not know what kinds of data they must obtain if the problem were to be solved.

This theoretical advance came in 1915, when Einstein propounded his general theory of relativity. For the first time, a theory existed that made it possible to develop mathematical equations describing the structure of the universe.

One sometimes hears references to "Einstein's theory of relativity." Such references can be misleading because there are, in fact, two relativity theories, which are quite different from one another. Einstein's first relativity theory, called the special theory of relativity and published in 1905, deals with the behavior of objects that travel at velocities approaching the speed of light. It is also the theory in which the famous equation $E = mc^2$ is developed. The general theory of relativity, which appeared ten years later, is Einstein's theory of gravitation. It deals with gravitating bodies and their effects on one another and on the space they occupy.

Although the general theory is based on fairly simple assumptions, it is mathematically very complicated. In most cases, the equations that it produces are so complex that they cannot be solved. As a result, even though Newton's law of gravity has supposedly been replaced by Einstein's, scientists still use Newton's much simpler equation whenever they can. In most cases, this is a perfectly good approximation. At the surface of the earth, for example, Newton's law of gravity gives results that have an accuracy of better than one part in a billion. Newton's law is also perfectly adequate for calculating the trajectory of a space vehicle traveling to the planet Jupiter, to cite just one example. Under such circumstances, it would be foolish to attempt to use Einstein's more complicated equations.

When one deals with the entire universe, on the other hand, it is necessary to use general relativity. Fortunately, the universe is, in a certain important sense, a simple thing. One could not use Einstein's equations to precisely describe the motions of several bodies that were revolving around one another; even Newton's theory would have trouble doing that. However, when one talks of everything that exists, one can make simplifying assumptions. Simple questions exist that describe the behavior of the universe as a whole.

The game of bridge furnishes a useful analogy. The set of thirteen cards held by a player can be hard to bid and difficult to play. Numerous books have been written about the problems that are likely to arise. But the structure of the entire deck of cards with which bridge is played is comparatively simple. There are thirteen denominations, each of which is repeated four times (as spades, hearts, diamonds, and clubs). The structure of the deck can be summed up in a sentence.

The Structure of the Universe

Einstein himself was the first to use his general theory of relativity to describe the structure of the universe. In 1917, two years after he propounded the general theory, Einstein published a paper describing his conception of the cosmos. He began by assuming that the universe was static, that it neither expanded nor contracted. At the time this seemed a reasonable assumption to make.

Philosophers and scientists alike had always conceived of the universe as an unchanging thing.

Einstein found, however, that his equations did not imply a static universe. At least, this was not the result that he obtained when he used the equations in their simplest form. However, on further reflection this did not pose a serious problem. There was an easy way around it: Einstein added an arbitrary quantity to his equations, calling it the cosmological constant. He found that if he did this, a static universe would result.

Inventing the constant may seem like fudging, but it was a perfectly legitimate mathematical technique, one that makes equations more general. (In mathematics, the cosmological constant is a constant of integration.) And if empirical data later reveal that the constant that has been added is not really needed, it can easily be eliminated again by setting it equal to zero.

There was one problem, however. Einstein's constant seemed to have a peculiar character: If it wasn't precisely zero, then there had to be an odd antigravity force operating throughout the universe. In physical terms, putting in the constant amounted to adding a quantity that would counteract the force of gravity at large distances. Einstein realized that no such force had ever been observed. But it had to be present, he reasoned. The only alternative was to conclude that the universe had to be either expanding or contracting.

According to Einstein's paper, ours was a static universe in which the gravitational forces that drew matter together were exactly balanced by a cosmological force that held them apart. However—and Einstein should have seen this—such a perfect balance could never be maintained. For example, if the universe happened to expand by an infinitesimal amount, then that expansion would inevitably continue. After all, when the universe expands, galaxies move farther apart. When they do, the gravitational forces between them become weaker. And if gravitational attraction became weaker, the repulsive force would dominate, and the expansion would continue at an ever increasing rate. On the other hand, if such a universe contracted by a small amount, then the contraction would inevitably continue. In this case, the forces of gravitational attraction would increase, and the cosmological repulsion would no longer keep galaxies apart.

Einstein's universe, in other words, was somewhat analogous to that pencil balanced on end. There is no particular reason for the pencil to fall one way or another; forces may initially be perfectly balanced. However, such a delicate balance cannot be maintained, since the smallest disturbance will cause the pencil to topple one way or another.

By now it should be obvious that Einstein's 1917 paper on the structure of the universe was not his most brilliant piece of work. In fact, it contained a number of mistakes, and it wasn't long before someone pointed them out.

In 1922, the Russian mathematician Alexander Friedmann showed that, at one point in the paper, Einstein made an error that high school freshmen are warned about when they begin to study algebra. He had divided by a quantity that was sometimes equal to zero. Division by zero is forbidden in algebra because it can lead to paradox. It is generally one of the elements of those deceptive "proofs" that purport to show that 0 = 1 or that 1 = 2.

When Friedmann corrected this error, he discovered that Einstein's "static" universe wasn't really static at all. He found that the universe was like that pencil balanced on its end. And, of course, if the universe had a tendency to go into a state of expansion or contraction anyway, the cosmological constant was not needed. There was no reason to put it into the equations if it didn't even accomplish the purpose for which it was intended.

When Einstein saw what Friedmann had done, he initially protested. But he finally had to admit that he had made mistakes and that the Russian's corrections to his theory were valid. He wasn't sure that he liked the idea of a universe that was expanding or contracting, but there seemed to be no way to avoid it.

When Einstein made his remark about "the greatest blunder of my life," which I quoted in Chapter 1, he was referring to the cosmological constant. Today, scientists tend to think that Einstein's paper was not quite as incompetent as he believed it to be. It is true that Friedmann was correct when he demonstrated that a static universe was impossible. On the other hand, there is a possibility that the cosmological constant could be real. As we shall see later on, there are reasons for thinking that it was quite large for a short period of time when the universe was only a fraction of a second old. And small remnants of this antigravity force may still operate in the universe today.

CURVED SPACE

Not only does Einstein's general theory of relativity describe the expansion of the universe, but it also has a great deal to say about the structure of the universe. The theory does not tell us whether the universe is infinite or finite; this is something which must be determined empirically. But it does relate the infinity or finitude of the universe to quantities that can be measured, such as the average density of matter in the universe, and the rate at which the expansion is being slowed down by the retarding force of gravity.

Einstein's general theory of relativity introduced the concept of curved space. By now, this is probably one of the most familiar ideas of modern physics. It has been explained in numerous books and on many educational television programs. But despite its familiarity, a short explanation is certainly in order, because curved space is perhaps one of the most significant conceptions in modern physics. Without some understanding of it, we cannot understand what our universe is like, or why it has evolved in the manner it has.

Newton thought of gravitational force as "action at a distance." He did not know why distant bodies should exert an attraction on one another. He only knew that one had to assume that they did if one were to explain the motions of planets and other astronomical bodies.

Einstein substituted the concept of curved space for this mysterious "action." According to Einstein, every massive body changes the geometry of space in its vicinity. This effect is sometimes illustrated by the analogy of placing weights on a rubber sheet. The resulting depressions are said to correspond to regions where gravity is strongest.

One can also use the analogy of a roulette wheel. When a roulette ball is set in motion, it moves in a circle because it is constrained by the curved geometry of the wheel. It travels in a groove at the rim. Then, as it loses energy, it begins to fall toward the center. And, of course, if the ball were not constrained, it would go flying across the casino when it was given its initial impetus.

The other way to describe Einsteinian curved space is to talk about its geometry, which is somewhat different from the Euclidian geometry that we learn in high school. There is a theorem in Euclidean geometry, for example, which states that the sum of the angles in a triangle is always exactly equal to 180°. In the non-Euclidian geometries that describe curved space, on the other hand, the sum may be more or less than that figure. (See Figure 1.)

Unfortunately, one can't test the general theory of relativity by going out and measuring the angles of triangles. This will not be possible until some day far in the future (if such a day ever comes) when observations can be made from points that are many light years apart. The deviations from Euclidian geometry that one would find on the surface of the earth are simply too small to be measured.

Interestingly, an attempt at this kind of measurement was actually made as long ago as 1827 by the German mathematician Karl Friedrich Gauss. Gauss measured the angles of the triangle formed by the mountain peaks called Brocken, Hohehagen, and Inselberg. As one might expect, the results were inconclusive. No, Gauss had not anticipated Einstein's theory, but had realized that non-Euclidian geometries were possible, and he thought he might be able to determine experimentally whether the geometry of space was Euclidian or not.

One should not conclude, however, that all the effects predicted by general relativity are too small to measure. As a matter of fact, its conclusions regarding the effects of the curvature of space have been confirmed by experiment. During the 1960s and 1970s, a great number of precise experiments were performed, and the theory passed its tests with flying colors.

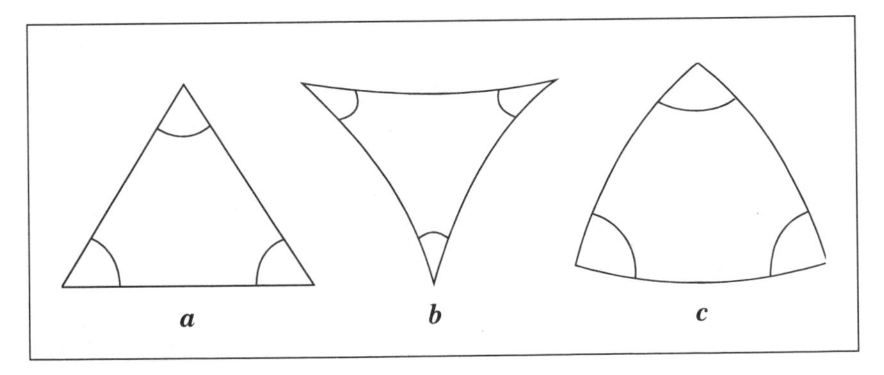

Figure 1 *In flat space the sum of the angles of a triangle is always 180°* ***(a)****. But in curved space, the sum of the angles can be either less than 180°* ***(b)*** *or greater than 180°* ***(c)****.*

Some of the experiments involved bouncing laser beams off pieces of reflecting material that astronauts had placed on the moon. Others confirmed that, just as Einstein had predicted, time slowed down in a gravitational field. This effect, incidentally, is the analogue of the curvature of space. One could say that a gravitating body "curves" time by slowing its progress very slightly while it simultaneously causes the space in its vicinity to bend. The effects on time are not large, but they are easier to measure than spatial curvature because scientists can make use of very accurate atomic clocks. An example of such a clock is the isotope cesium 133, which emits microwaves at a frequency of 9,192,631,770 cycles per second. It has been shown that such a clock will "tick" at a slower rate if gravitational forces become stronger, for example, if it is placed in a position nearer to the center of the earth.

One of the most dramatic confirmations of the general theory of relativity is provided by observations of the gravitational lens effect (see Figure 2). Astronomers can sometimes observe the same quasar at two or more different places in the sky. Quasars are the brightly glowing cores of young galaxies, and they are located billions of light years away. Now, if a galaxy happens to be located

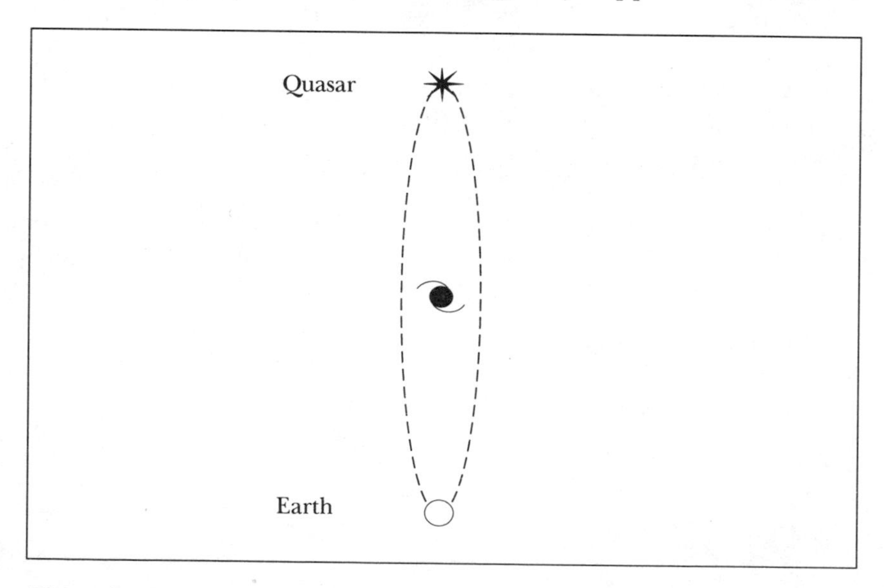

Figure 2 *When a galaxy is situated between a distant quasar and the earth, the light from the quasar can bend around opposite sides of the galaxy, producing multiple images of the quasar. This is called the gravitational lens effect. The magnitude of the bending of the light rays is exaggerated for greater clarity.*

between a distant quasar and the earth, the path of a light ray coming from the quasar will be bent by the galaxy's gravity. A body such as a planet does not travel in a straight line when it is near a gravitating body, and a light ray does not either. Naturally, the deviation of the light ray will be much less: One does not see light traveling in circular or elliptical orbits, for example. However, the bending of the path followed by a ray of light is substantial enough to be observed.

If two light rays coming from a quasar graze opposite sides of a galaxy, their paths may be bent in such a way that both converge on the earth. In such a case, a terrestrial astronomer will see two different images of the same quasar on opposite sides of the galaxy. Actually, the general theory of relativity predicts that not two but an odd number of images should be observed in such a case. This prediction turns out to be correct: Three quasar images are sometimes seen. But astronomers will often observe just two because the third is too faint to be seen.

The Closed Universe

According to the general theory of relativity, the overall structure of the universe is related to the quantity of matter that it contains. If the average density of matter in the universe is greater than a certain critical value, then gravity will cause space to close back upon itself, and the universe will be finite. The critical density is not very great—roughly equal to one hydrogen atom per cubic meter, or about 1 ounce for every 50 million billion cubic miles. Of course, the universe is an enormously big place, and it should be obvious that even such a low density as this corresponds to a great deal of matter.

A closed universe is finite, but it has no boundaries. Space in such a universe is said to be positively curved. This positive curvature is the three-dimensional analogue of the two-dimensional surface of the earth. One never comes to the "edge" of the earth. There is really only one fundamental difference between the two cases. It is possible to circumnavigate the earth. For example, if an airplane flies for long enough in any direction, it will eventually return to its starting point. This cannot happen in a closed universe. Even if there were beings who lived for many billions of years and who possessed spaceships with unlimited range, it would still not be possible to travel all the way around the universe.

The problem is not that the universe is too big, but rather that a closed universe does not last long enough for anyone to circumnavigate it. Even a light ray emitted in the big bang fireball could not accomplish this before the universe collapsed. And a closed universe must inevitably collapse. If the density of matter is greater than the critical value, then the gravitational retarding forces exerted on the matter in the expanding universe are great enough that the expansion must inevitably halt. In the case of our universe, the expansion has been slowing for the last 15 billion years (again, I am assuming a 15-billion-year age for the universe). In a closed universe, the expansion will slow in this manner and eventually stop. When that happens, gravity will continue to have an effect, and a phase of contraction will begin. Finally, after many billions of years, the universe will be destroyed in a big crunch that is the analogue of the big bang.

SPACETIME

Before I describe other possible kinds of universes, I would like to comment on some technical points. The first of these has to do with the often-used term *spacetime.* Mathematically, general relativity is a four-dimensional theory. It deals with three dimensions of space and one of time. There is nothing very mysterious about this; after all, Newton's laws of gravitation and motion are also four-dimensional. They also describe a world with three dimensions of space and one of time. The only real difference is that, in relativity, space and time become bound up with one another in a way that they do not in the older, classical physics. I referred to one such effect when I pointed out that, when a gravitating body curves space, it also slows down time; the two effects are inseparable.

The next important point is that, when one compares the curved space of a closed universe to the curved surface of the earth, one is not implying that there are any extra spatial dimensions. The earth's surface curves in a third dimension of space. Curved space does not bend in a fourth dimension; there is no fourth dimension of space in relativity. Similarly, though one can fly above the surface of the earth, it is not possible to "fly up" in the universe as a whole in any extra dimension and view the universe from above or from "outside." This is one of those cases where an analogy becomes misleading if it is pursued too far.

It is useless to try to visualize the appearance of a closed universe. Not even mathematical physicists can do that. However, they can describe the curvature of space in a closed universe by means of mathematical equations. And, given the numerous experimental confirmations of general relativity that exist today, there is every reason to think that these equations are correct.

It is also important to know that Einstein himself seems to have preferred the idea of a closed universe because he found the concept more aesthetically appealing than the alternatives. Certain contemporary scientists, such as the British physicist Stephen Hawking, also appear to like the idea of a closed universe. But the only way to determine if the universe really is closed is to try to answer the question empirically. In principle, this could be done—one could measure the average density of matter, for example. However, as we shall see, this turns out to be an enormously difficult task in practice.

OPEN AND FLAT UNIVERSES

If the average density of matter is less than the critical value, then the universe is said to be open, and space is said to be negatively curved. Comparisons are sometimes made between this kind of spatial geometry and the curvature of a saddle. But a relatively more important point is the fact that, under such circumstances, space does not curve back upon itself. An open universe is infinite. In such a universe, moreover, the expansion never comes to a halt. (See Figure 3.)

One can think of an open universe as one in which gravity is not strong enough to cause space to close in upon itself. In such a situation, gravitational forces will not be strong enough to halt the expansion. If the density of matter is very close to the critical value, the rate of expansion may diminish appreciably over billions of years. The expansion rate slows down in an open universe just as it does in one that is closed. However, as the galaxies move apart, the gravitational retarding force becomes progressively weaker. In an open universe, the point is eventually reached

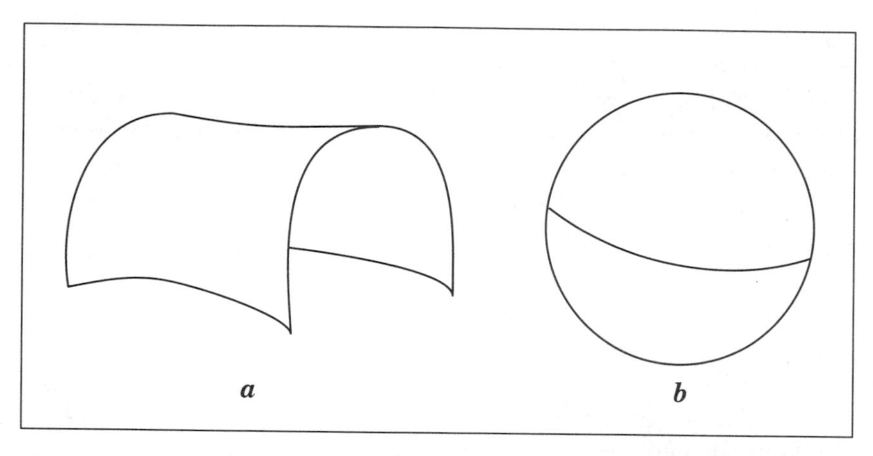

Figure 3 *Curved three-dimensional space cannot be pictured. But it is possible to draw two-dimensional analogues. In diagram* ***(a)****, a two-dimensional sheet has been bent into a kind of saddle shape. It does not close upon itself. Like a flat plane, it extends out to infinity. In diagram* ***(b)****, a two-dimensional surface has been molded into a sphere. It does close upon itself. If an object moved in any direction, it would eventually come back to its starting point.*

where the gravitational retardation is no longer significant because the galaxies are too far apart. Once this point is reached, an open universe will go on "coasting" forever.

An open universe can be said to be expanding even though it is infinite in extent. As I pointed out in the previous chapter, expansion is defined as the tendency of the matter in the universe to become more dispersed. A closed universe does become larger; it has a definite, finite volume. An open, infinite, universe doesn't, but it can still expand.

There is also a third possible configuration that our universe could have: It could be flat. A flat universe is one in which the density of matter is exactly equal to the critical value. One can think of a flat universe as a kind of borderline case between open and closed. In such a universe, the average curvature of space is zero. Such a universe is also infinite. In fact, it corresponds pretty closely to Newton's conception of an infinite cosmos. The only significant difference is that Newton did not conceive of the universe as expanding.

Like an open universe, a flat universe would go on expanding forever. However, since the gravitational retarding forces are greater than they are in an open universe, the expansion would slow down faster. One can think of a flat universe as one in which the expansion grows continually slower and slower and slower but never quite reaches zero. (See Figure 4.)

The idea of a flat universe is, however, an idealization that does not correspond to anything that could exist in reality. If the actual density of the universe were greater than the critical value by one part in a trillion, then the universe would be closed. And if it were less by a factor of one part in a trillion, the universe would be open. Even if the deviation from the critical value were much less than, say, a trillionth of a trillionth of a trillionth, the universe would still be either open or closed. Nevertheless, the idea of a flat universe is a useful concept. As we shall see in Chapter 4, many scientists think there are good reasons to believe that our universe is very nearly flat. They point out that there are theoretical reasons for believing that the density of our universe may be so close to the critical value that we will never know whether it is infinite or finite,

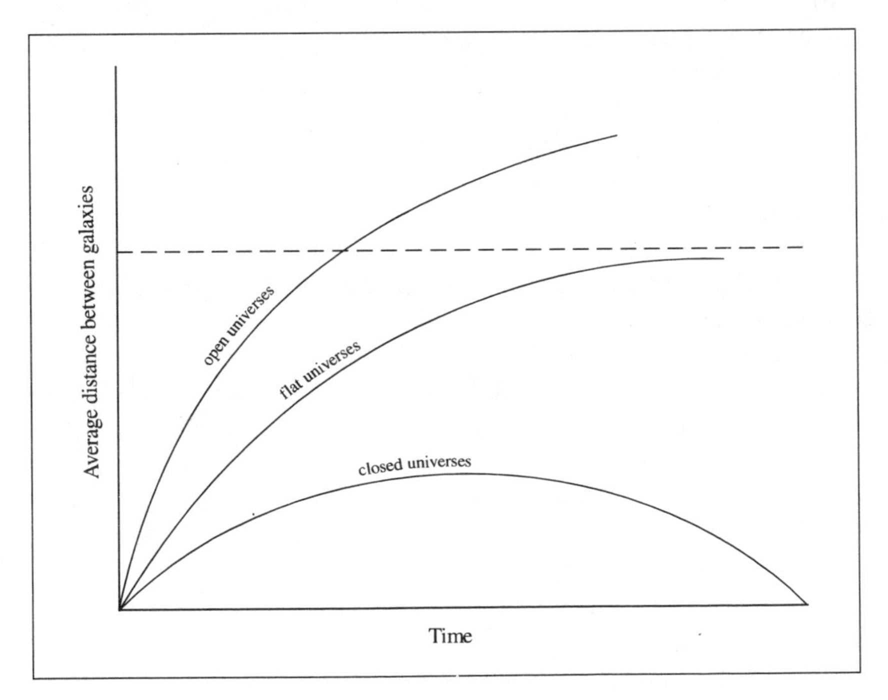

Figure 4 *This diagram depicts the expansion of open, flat, and closed universes. An open universe will keep on expanding forever, whereas a closed universe will eventually enter a phase of contraction and collapse upon itself. Finally, a flat universe will expand ever more slowly and approach a certain limiting size.*

whether it is destined to expand forever, or whether it will eventually be destroyed in a big crunch.

MEASURING THE DENSITY

Theoretical arguments, such as the one we shall meet in Chapter 4 concerning what the density of the universe is expected to be, can be very useful. But they are no substitutes for observation and experiment. Therefore, if one wants to know whether the universe we live in is open, closed, or nearly flat, one should attempt to settle the matter empirically.

In principle, there are two main methods of doing this. One can attempt to measure the matter density of the universe directly, or one can attempt to determine how fast the expansion of the universe is slowing down. After all, the slowing of the expansion and the density of matter are closely related quantities. The greater the amount of matter that is present, the stronger the gravitational retarding forces will be. If the slowing of the expansion could be measured, calculation of the average density of matter would be quite a simple affair.

Unfortunately, both methods turn out to be very inaccurate. Astronomical observations fail to give us a precise estimate of how much matter the universe contains. At best, they only make it obvious that the universe contains a great deal of matter that scientists cannot see.

It is really not very difficult to estimate the number of stars in a galaxy or their total mass. Both of these can be accurately determined by measuring the quantity of light that a galaxy emits. As soon as astronomers determine how bright a galaxy is, they immediately have a pretty good idea of how much the stars in the galaxy weigh. By measuring the quantity of light, they can determine how many stars a galaxy contains.

Nor is it hard to estimate the mass of interstellar and intergalactic gas. Cool gas emits radio waves, which can be detected by radio telescopes on the surface of the earth. Hot gas, on the other band, gives off x-rays. These x-rays tend to be screened out by the earth's atmosphere, but they can be observed with instruments placed on rockets and satellites.

The contributions that stars and gas clouds make to the average mass density of the universe have, in fact, been calculated, but it turns out that these objects account for no more than a few percent of the quantity of mass that is needed to close the universe. If stars and gas were all that existed, the density of the universe would be much less than the critical value.

DARK MATTER

Astronomers know, however, that other matter is present. Though they can't see it, they can observe its effects. As long ago as 1932, the Dutch astronomer Jan Oort concluded that some kind of dark matter existed. He observed the motions of certain stars in our galaxy and concluded that these motions could not be explained if one assumed that the visible matter was the only mass present. The gravitational forces exerted by such visible matter were simply too weak to cause all the motion observed.

Oort calculated that about 50 percent of the mass in the disk of our galaxy resided in the matter (stars and gas clouds) that was visibly present. The other 50 percent was due to something that could not be seen. Oort suggested that the discrepancy might be caused by the presence of small stars that were too faint to be observed. Contemporary astronomers discount this idea, however. They have observed that such stars simply do not exist in sufficient numbers to cause the observed effects. Whatever it is that causes these motions is truly dark.

When scientists speak of "dark" matter, they are not referring to material that is dark in color like a dust cloud or a piece of charcoal. They mean that it emits no light or any other kind of radiation that can be detected, and therefore it cannot be seen at all. The dark matter in our galaxy is not some dark light-absorbing material such as interstellar dust. Interstellar dust can be easily observed because it blocks out light from stars that are behind it.

In 1933, the year after Oort published his results, the California Institute of Technology astronomer Fritz Zwicky observed a large cluster of galaxies in the constellation Coma Berenices. He noted that, even though the galaxies in the cluster were apparently held together by their mutual gravitational attraction, the mass that was present in the galaxies' stars accounted for only a fraction of the quantity that was needed. According to Zwicky, there was a "missing mass" problem.

The term *missing mass* is no longer used today. After all, nothing is really missing. The dark matter is obviously there. The only problem is that, since it gives off no radiation that scientists can study, they have no way of determining what it is.

There has been no lack of suggestions as to what the dark matter might be. Dim, faintly glowing stars might account for some of it. Even though these do not exist in numbers sufficient to account for the motions that Oort observed, there are probably some present.

Some dark matter could exist in the form of black holes. Black holes are remnants of dead stars that have gravitational fields so strong that nothing—not even light—can escape from them. Scientists do not really have any good way of estimating how many black holes a typical galaxy might contain.

Another reasonable suggestion is that some of the dark matter might exist in the form of bodies about the size of the planet Jupiter. Jupiter is quite a massive planet. If there were a lot of bodies like it in our galaxy, they could make a significant contribution to the total mass. But since they would emit no light, they could not be seen through telescopes.

As I will explain later, it is possible to determine the quantity of dark matter that is present in a galaxy, and it is also possible to estimate the quantity of dark matter present in a cluster of galaxies. The quantity can be inferred in both cases from its gravitational effects. This is what Zwicky did when he observed the cluster in Coma Berenices. However, it is impossible to determine precisely how much dark matter is present in the vast spaces between clusters. And since our universe is mostly empty space, the problem of measuring the mass density of the universe is a difficult one indeed.

If one depends on astronomical observations alone, it is only possible to set a lower limit on the average mass density of the universe. One can conclude that, at the very least, the density is a few percent of the critical value. But of course it could be much, much greater than that. How much greater is impossible to know with certainty.

However, we don't necessarily have to give up on the idea of determining what the mass density of the universe might be. Indirect methods for obtaining a good estimate are available. For example, information about the mass density of the universe can be obtained from measurements of the concentrations of deuterium and other light elements in the universe.

The amount of deuterium that is present now is related to the density of the universe at the epoch in which it was formed. The denser the universe was then, the greater the number of deuterium-creating reactions that could have taken place. After all, if a lot of protons and neutrons are confined to a small space (and deuterium is created when a proton and neutron combine), they will wind up sticking to one another much more often than they will if they are so dispersed that they rarely meet. Measuring the quantity of deuterium that is present today thus provides us with information about the quantities of hydrogen and helium that existed when the universe was young. And if one knows how dense the universe was then, it should be possible to estimate how dense it is now.

When the calculations involving deuterium and other such light isotopes as helium 3 and lithium 7 are performed, one finds that the density of the universe is about 6 percent of the critical value. Thus it would appear that the universe must be open. At least that is the conclusion one could draw were it not for a big hole in this argument.

When this calculation was first done in 1974, scientists did in fact conclude that we lived in an open universe. Today, however, they are not so sure, for they have realized that this calculation sets limits only on the amount of baryonic matter that can be present. It says nothing about other kinds of dark matter that might conceivably exist.

Baryonic matter is matter that is made up of such common particles as protons, neutrons, and electrons. It is called "baryonic" because protons and neutrons belong to a class of particles that physicists call baryons. The electron is not a baryon, but it can be left out of the accounting because it is so light that it does not contribute significantly to mass (a proton is 1836 times heavier than an electron, for example). If one wished, one could replace the term *baryonic matter* with *ordinary matter*. After all, stars, planets, and interstellar dust and gas are all made of baryons and electrons, as are such things as automobiles, hamburgers, dogs, computer chips, bacteria, and human beings.

If baryons and electrons were the only particles that existed, one could indeed conclude that the universe was open. However, physicists have discovered the existence of numerous other subatomic particles in the laboratory, and they have suggested on theoretical grounds that there might be other kinds of particles that have not yet been observed. No one knows how many of these particles might be roaming the vast intergalactic spaces that make up the greater part of our universe. Even in relatively small numbers they would contribute significantly to the overall mass density.

One must conclude that attempts to determine the mass density of the universe by empirical methods have not been terribly successful. We can say that, if the greater part of the matter in the universe is baryonic, the universe must be open; but no one knows just how much nonbaryonic matter there might be. We have not been able to determine, therefore, whether the universe is open, closed, or flat.

Much more can be said on the subject of dark matter. In fact, it is a topic that will come up on a number of different occasions during the course of this book. At the moment, however, we are reviewing the empirical evidence that might have some bearing on whether the universe is open or closed, and there is one more topic to be covered.

THE SLOWING OF THE EXPANSION

In principle, one could calculate the mass density of the universe if one could determine the rate at which the expansion is slowing down. And there is no compelling reason why this could not be done, if astronomers could only make measurements that were accurate enough.

To be sure, it is impossible to measure the rate of expansion today, wait a hundred million or a billion years and then measure it again. Fortunately, it is not necessary to do anything like this: Astronomers have a technique that should work just as well. All they need to do is look back in time and determine how much faster the galaxies were moving away from one another in the past.

As I have pointed out previously, seeing into the past is not a particularly difficult task. One need only observe galaxies that are very far away. If one looks at a galaxy at a distance of a billion light years, one sees it as it appeared a billion years ago.

Unfortunately, it is somewhat more difficult to determine how rapidly the universe was expanding at past epochs. For that matter, astronomers have not even succeeded in determining exactly how rapidly the universe is expanding today. The problem is that, to determine the rate of expansion, one must measure the distances of the galaxies from the earth. And measuring these distances is one of the most difficult tasks in science.

In order to calculate the expansion rate, one must know two things: How far away the galaxies are, and how fast they are receding. The latter quantity can be determined very accurately by measuring redshifts. Distances, however, are known only very approximately. In fact, different scientists use different distance scales for measuring the universe. One of these scales is twice the size of the other.

The first step is measuring the distances to stars in our own galaxy. This can be done quite accurately by the method of parallax (see Figure 5). If an astronomer wants to know how far away a star is, he or she determines its position, and then determines its position again six months later when the earth is on the other side of its orbit. Since the earth will have shifted its position by nearly 300 million kilometers (or 186 million miles—twice the distance to the sun), the position of the star will appear to have shifted. Its distance can then be found by triangulation.

Unfortunately, this method cannot be used to measure the distances of other galaxies. In fact, it gives the distance of only the nearest stars in our own galaxy, since it is accurate only to distances of about 100 light years, which is about one-tenth of 1 percent of the diameter of the Milky Way. The galaxies that astronomers observe, on the other hand, may be millions or billions of light years away.

A number of alternative methods can be used to estimate intergalactic distance. They are all based on the idea that if one knows the intrinsic brightness of an object, one can tell how far away it is by measuring its apparent luminosity. For example, a 100-watt light bulb will look very bright if it is a few feet away, and a searchlight will appear dim if it is distant enough.

The apparent brightness of galaxies can be measured, but the distance estimates thereby obtained are not very accurate. One of the problems is that mapping out distances is a step-by-step process. One method is used to measure the distances of nearby galaxies, and a second is used to estimate the distances of those

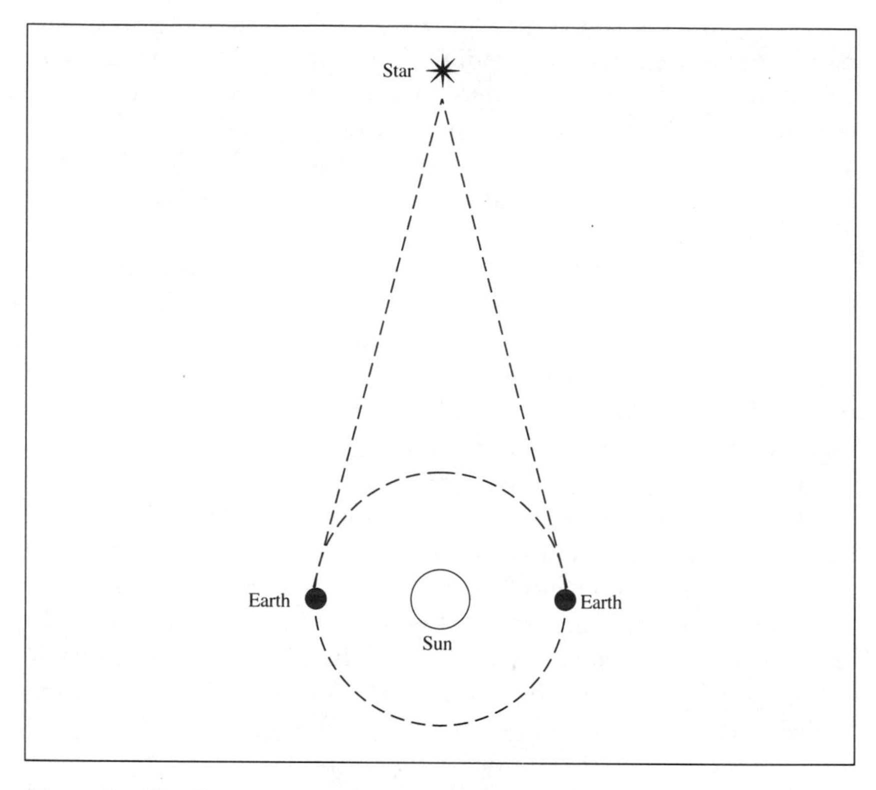

Figure 5 *The distance to nearby stars can be found by the method of stellar parallax. The apparent direction of the star is measured at times 6 months apart, when the earth is on opposite sides of its orbit. The distance can then be found by triangulation. The diagram is not drawn to scale. In reality a star would be much, much farther away than it appears here.*

farther away. Then one switches to a third method, and so on. Although some of these methods are reasonably accurate, all contain uncertainties, and the errors accumulate as one progresses from one step of the ladder to the next.

The distances of the nearest galaxies, such as the great galaxy in Andromeda, are determined by measuring the brightness of certain characteristic stars that they contain. The intrinsic brightness of a certain class of stars, called Cepheid variables, is known. The accuracy of this method is not as great as that of the parallax procedure, and the error it introduces is carried over when these distance estimates are used as the basis for calculating the distances of galaxies further out, in which individual stars cannot be observed. And the farther out one goes, the more inaccurate the results become.

When astronomers examine galaxies that lie at very great distances, hundreds of millions or billions of light years, for example, they must use the galaxies themselves as distance indicators. Assuming that the brightest galaxies in different clusters have about the same intrinsic luminosity, they measure the apparent brightness of the brightest galaxy in order to determine the distance of the cluster from the earth. This isn't as bad a way of measuring distance as it may sound, because the luminosity of very bright galaxies is more uniform than one might suspect. In fact, 90 percent of them have brightnesses that vary by less than a factor of 2.

On the other hand, this procedure is far from precise, and additional complications raise further questions about the estimates it provides. The most distant galaxies that astronomers observe are more than 10 billion light years away. Obviously, a lot of things can happen in 10 billion years, and one must try to estimate how the brightness of galaxies today compares with the brightness of 10 billion years ago.

As eons of time pass, stars grow old and die. If this were all that happened, galaxies would gradually grow dimmer. However, other things are going on as well. Galaxies often collide, and a large galaxy may swallow a smaller one. When this takes place, a galaxy becomes brighter, for the simple reason that it now contains more stars. Because of the various uncertainties involved, astronomers are not quite sure how the brightness of the average galaxy changes in time.

As if matters were not difficult enough already, astronomers have also discovered that the expansion of the universe is not completely uniform. The universe seems to be full of streaming motions that cause entire groups of galaxies to move in one direction or another relative to the universe as a whole. For example, the group of galaxies containing the Milky Way seems to be moving through space at a velocity of about 600 kilometers per second. It is apparently in the gravitational grip of some huge mass that lies hundreds of millions of light years away. The universe, in other words, is churning with motion, and this makes it more difficult to determine how rapid the average expansion is.

The numerous uncertainties about the distance scale are responsible for the uncertainty about the age of the universe. The rate of expansion of the universe has only been determined to within a factor of 2. Some scientists prefer a lower figure, while others maintain that the expansion is progressing twice as fast.

Obviously, if one does not know exactly how rapidly the universe is expanding, it is impossible to calculate when the expansion began. This is why we do not know whether the age of the universe is 10 billion years, or 20 billion, or something in between. Naturally, those uncertainties cause uncertainty about the rate at which the expansion is slowing down as well.

BUT THINGS AREN'T ALL THAT BAD

We may begin to feel overwhelmed by uncertainty, but matters are not quite as bad as one might think. Though it is impossible to tell for sure whether a cluster of galaxies is, say, 4 billion or 8 billion light years away, relative distances can often be quite accurately determined. Astronomers can be reasonably confident about their figures when they say, for example, that galaxy A is 4.3 times as far away as galaxy B. Similarly, they can make definite statements of this sort: "The light from this quasar was emitted when the universe was one-tenth its present age." Astronomers cannot be certain whether the quasar emitted its light 1 billion years after the big bang, or 2 billion, but they can rely on relative figures like that "one-tenth."

The determination of such relative quantities helps determine how fast the expansion of the universe is slowing down. But it doesn't help enough to allow astronomers to draw any real conclusions about the biggest questions explored in this chapter. They can say that the expansion seems to have slowed somewhat over the past few billion years. But the data that have been accumulated are consistent both with the idea that the universe is open and with the idea that it may be closed.

Although astronomers have been trying to solve the problem of the nature of the universe for decades, it is so difficult that they have made little real progress. At best, we are now able to say that the density of the universe is probably somewhere between one-tenth and ten times the critical value. If one depends on empirical methods alone, on which this measurement is based, we must conclude that the feature of the universe many would consider to be the most important has been described to within a factor of 100.

3

When Did Time Begin?

Scientists who try to understand the nature and evolution of the universe find themselves in a somewhat paradoxical situation. Although they can see nearly all the way back to the beginning of time, they aren't sure when the beginning of time was. When they observe the cosmic microwave background, they are looking at the universe as it appeared 300,000 years after the big bang. And yet they aren't sure whether 10 billion years have elapsed since then, or whether the true figure is 15 or 20 billion.

One way in which scientists might finally determine when exactly the universe began is by discovering what conditions existed during the first moments. For example, most of the helium that exists today, and probably all of the deuterium, was manufactured when the universe was two or three minutes old. By measuring the abundances of these substances, scientists can obtain data that allow them to make deductions about the conditions of the universe when it was very young indeed.

Furthermore, there is a very real possibility that scientists will eventually be able to "see" back into the very moments after the big bang. Theoretical calculations indicate that when the universe was only a fraction of a second old, various kinds of exotic subatomic particles should have been created. If these particles are eventually observed, it will be possible to make deductions about the exact state of the universe at that fraction of a second into its existence.

It has been said that the early universe is a laboratory for particle physics. What scientists mean by this observation is that the energies existed in the early universe that far exceed any that can be attained in terrestrial particle accelerators today, and important information about the microworld that cannot be obtained by ordinary lab experiments may be inferred from astronomical observations. Similarly, cosmology is enriched by discoveries made in terrestrial laboratories by physicists. For example, every time a new particle is discovered and its properties studied, astronomers learn a little more about the possible character of dark matter.

What kinds of particles could have been created by the high energies in the very early universe? There are numerous different theoretical possibilities, so many, in fact, that these particles are generally grouped together under their family name and referred to as WIMPs (weakly interacting massive particles). As I write this, not a single WIMP has ever been detected, either in cosmic rays or in experiments conducted in particle accelerators. So as yet there is no experimental evidence that they exist. However, many scientists are convinced that one or more varieties of WIMP will eventually be observed. If so, they will gain a much clearer picture of the conditions in the very early universe, and they may be able to clear up a number of the most perplexing cosmological problems.

16-BILLION-YEAR-OLD STARS IN A 10-BILLION-YEAR-OLD UNIVERSE?

It is all very well to explore the nature of the universe when it was 300,000 years, or two minutes, or a thousandth of a second old. However, such explorations do little to answer the question that was posed in the chapter title: Exactly when did time begin?

Even though the age of the universe is uncertain, many clues exist that tell us something about how old it might be. Examining these clues can lead to conflicting results, but whenever evidence seems to be contradictory, that is often the best reason to take a long look at it.

One of the puzzles has to do with the age of certain very old stars. Like many other galaxies, the Milky Way is surrounded by compact groups of stars called globular clusters. These clusters, each of which contains about a million stars, revolve around our galaxy in a manner similar to that in which the planets revolve around the sun. Naturally there are differences. The clusters tend to have very elongated orbits, and there are about 200 of them, as compared to nine planets in our solar system. However, the principle is the same.

The stars that make up the globular clusters are very old. Many of them appear to have been formed about 15 billion years ago. Scientists used to generally believe that those stars were much older than the galaxy itself. Until 1992 the oldest stars that had been found in the Milky Way were white dwarfs that were only a little more than 9 billion years old.

But then astronomer Young-Wook Lee of Yale University announced that he had discovered some stars in the central bulge of our galaxy that were about 1.3 billion years older than the old stars in globular clusters. Lee had peered through a small gap in the clouds of gas and dust that veil the galaxy's core and had spotted some stars of a type known as RR Lyrae variables. Analyzing the stars' color, he found that some of them were unusually blue. The only plausible way to explain this was to assume that the stars must be very old. A very old star would shrink as its nuclear fuel diminished. This would heat its surface layers and cause the light that it emitted to become bluer. In fact, the stars that Lee observed were much bluer than RR Lyrae stars of similar size that had been seen elsewhere.

This finding caused a certain amount of controversy within the astronomical community because it challenged the theories that the globular clusters were created before our galaxy coalesced out of a primeval cloud of gas. As I write this, the controversy has not been settled. However, the evidence strongly suggests that there are stars in globular clusters that are 15 billion years old, and possibly stars in the galactic core that have an age of 16 billion years.

If we accept the estimate of 15 billion years for the age of globular cluster stars as being reasonably accurate, then the universe is at least 17 billion years old. We must add the 2 billion years to get the estimated age because we know stars were not formed immediately after the big bang. It was necessary for time to pass before conditions were right, and 2 billion years seems to be a reasonable minimum figure. Unless scientists are completely mistaken about the conditions that existed in the early universe, initially no regions had dense enough concentrations of primordial hydrogen and helium for star formation to occur. Gravity first had to condense some of the gas into galaxy-sized—or globular-cluster-sized—objects. Even then, star formation could not begin immediately, because the gas had to be compressed further. A star, after all, is nothing but a ball of hot gas that has been greatly condensed by gravity.

So there is good evidence that we live in a universe that is 17 or 18 or 20 billion years old. Yet many scientists insist that there are also good reasons (these will be discussed later) for believing that the universe is much younger, and that the big bang could not have happened much more than about 10 billion years ago.

Of course, researchers are not claiming that the globular cluster stars and possibly the RR Lyrae stars in the galactic core are older than the universe itself. They are simply expressing doubt that these stars are as old as they seem to be. Though they aren't sure exactly what is wrong with the line of reasoning that leads to star ages of 15 or 16 billion years, they think that there must be a flaw somewhere. As we shall see, there exists other evidence which favors a younger universe.

The fact that there are such substantial disagreements about the age of our oldest stars does not, however, mean that we shouldn't take the estimates seriously. On the contrary, it will be very instructive to look at the evidence supporting the 15-billion-year estimate for cluster stars in some detail. Doing this will not lead to any definite conclusions. However, the reader who follows the discussion is likely to come away with a better understanding of how scientists study objects that are thousands, or millions, or billions of light years away, as well as gain some insight into why some results can be so accurate while others are so uncertain. The first order of business is to understand the process by which stars are dated.

THE EVOLUTION OF STARS

Astronomers believe that they understand the stages in the lives of stars very well. There is every reason that they should: They have observed stars as they are being formed, stars in their middle age, and stars that are dying. To be sure, one cannot follow the life of an individual star from its birth to its death; this is a process that takes hundreds of millions or billions of years. However, it is not difficult to find stars that are very young and to compare them with others that are middle-aged or old or dying. After all, although star formation began many billions of years ago, it is still going on now.

During the greater part of its life, a star burns in a reasonably calm and steady manner. Any changes in the quantity of energy produced by its burning occur only very gradually. For example, it is believed that our sun was about 40 percent less luminous when the earth was formed 4.5 billion years ago than it is now, and that its diameter was 4 percent smaller. These changes are hardly dramatic for such a vast period of time. In fact, the change in the sun's energy output after a million years of evolution would be so small that it would hardly be measurable.

When stars begin to die, on the other hand, they undergo changes that are quite dramatic. Some of them literally blow up. Very large stars, those that have more than about three or four times the mass of our sun, end their lives in supernova explosions. When their nuclear fuel is exhausted, they experience a catastrophic collapse, which is followed by a kind of massive rebound that sends their matter flying outward into space. When such an explosion takes place, so much energy is released that, for a brief moment, the dying star is billions of times brighter than our sun. And then, after a few weeks—an infinitesimal period of time by astronomical standards—the light begins to dim and the supernova fades away.

The death of a star like our sun is not quite so dramatic, but it is catastrophic enough. For example, in about 5 billion years, when it will have reached an age of about 10 billion years, our sun will expand into what is known as a red giant. As its nuclear fuel begins to be exhausted, it will gobble up the remaining amount at an accelerated pace. During a period of about 600 million years, it will continuously expand until it engulfs the earth and possibly also the planet Mars. At the end of this expansion, it will have a diameter 400 times greater than it does now, and its light output will have increased by a factor of 10,000.

But after this, the sun's remaining nuclear fuel will not last for long in astronomer's time. After another 100 million years, it will gradually shrink and decrease in brightness until it becomes a dim white dwarf. White dwarf stars are quite common in our galaxy. They have completely exhausted their nuclear fuel and continue to glow only because of their residual heat, like glowing coals.

How does all of this relate to the dating of stars exactly? When astronomers study globular clusters, they find a lot of stars that are in the later stages of their evolution. Thus they are able to conclude that the clusters are quite old. Furthermore, they are able to compute with a fair degree of accuracy how old the globular clusters are. This is not such a difficult thing to do. Astronomers believe that they understand the characteristics of stars well enough that they can compute how massive a dying star is. And if they know how much a star weighs, they can calculate how much time has passed since it was created.

Determining the age of a star is a bit like calculating how far a car has been driven by measuring the fuel left in its gas tank. If one knows that the tank was originally full, one can measure the number of gallons used. If one also knows what kind of mileage the car gets, one can obtain a distance. For example, if 10 gallons have been used and the car gets 18 miles per gallon, one can deduce that it has been driven 180 miles.

Determining the mass of a star gives astronomers information about how much nuclear fuel it started with. Stars do lose some mass during their lifetimes, but it is not difficult to take this factor into account if necessary. Observations also give astronomers information about how much nuclear fuel remains. As a star ages, changes take place in its color and luminosity, and it is not difficult to pick out stars that are nearing the ends of their lives.

The only other factor to be determined is the "mileage" that the star gets, that is, the rate at which it consumes its nuclear fuel. In principle, it should not be difficult to calculate this. Astronomers believe that they understand the nuclear reactions which take place in stars quite well, and they have studied these same reactions in the laboratory to determine how fast they proceed.

THE PROTON-PROTON CHAIN AND THE CARBON-NITROGEN CYCLE

There are two basic kinds of nuclear reactions that take place in stellar interiors: The proton-proton chain and the carbon-nitrogen cycle. Both reactions convert four hydrogen nuclei (four protons because a proton and a hydrogen nucleus are the same thing) into one helium nucleus. They are similar to the reactions that take place in a hydrogen bomb explosion. Because a helium nucleus weighs a bit less than the four protons with which the reactions begin, a certain amount of matter becomes available during the process to be converted into energy. This conversion is described by Einstein's famous formula $E = mc^2$. Here, E stands for the amount of energy that is created, m is the quantity of mass that is converted, and c is the velocity of light. In metric units, the velocity of light is 300,000,000 (3×10^8) meters per second, so c^2 is 90,000,000,000,000,000 (9×10^{16}), which is a huge number. Thus it is obvious that a small amount of mass produces quite a large quantity of energy.

Our sun is currently converting matter into energy by means of the proton-proton chain. As it ages, however, its internal temperature will increase, and the changed conditions will cause the carbon-nitrogen cycle to play a more prominent role. In this reaction, carbon and nitrogen nuclei act similarly to catalysts in chemical reactions. Even though this reaction is quite different from the proton-proton chain and considerably more complicated, the net result is the same: Four hydrogen nuclei are converted into one of helium.

BEAUTIFUL THEORIES AND UGLY FACTS

As the nineteenth-century British biologist T. H. Huxley once observed, it sometimes happens that a beautiful theory is destroyed by ugly facts. A hypothesis may be clear and logical and compelling and may still turn out to be wrong. Thus, even though scientists believe that they understand the reactions that take place within stellar interiors quite well, they would still like to find ways to test their ideas experimentally. And, in this case, the obvious way to do that would be to study the behavior of our sun. After all, it is the nearest and most accessible star.

One might think that it would be impossible to look into the core of the sun to see what is going on there. After all, the light that comes to us from the sun is approximately 2 million years old. The sun is so large and so dense that it takes this long for light that is produced in the sun's core to make its way to the surface. The light rays that we see are able to emerge into space and to start on their journeys to the earth only after they have been scattered, absorbed, and reemitted by matter within the sun many times. Thus studying this light really provides more information about the composition of the sun's outer layers than it does about anything that is going on inside the core.

Of course, it is possible to measure the quantity of energy that the sun emits, and an examination of sunlight provides important information about the chemical elements that are present in the sun as well. However, scientists would still like to find a way to observe the sun's core directly.

As it happens, there is a way of doing this. The nuclear reactions that take place in the sun's interior produce enormous quantities of elusive particles called neutrinos. Neutrinos are particles with very small—possibly zero—mass which interact with other particles of matter only very rarely. In fact, a neutrino could pass through the length of a block of lead that stretched from the earth to the nearest star without interacting with a single atom.

Yet neutrinos can be detected. The end products of their interactions with matter can be measured. Their reluctance to interact with matter is compensated for by the fact that there are a lot of them. Neutrinos are, in fact, by far the most common particles in the universe. Trillions of them pass through our bodies every second (but you needn't worry about the radiation hazard; no more than five or six will interact with any of your constituent atoms during your lifetime).

The great penetrating power of the neutrino is the characteristic that makes it an ideal probe for looking into the core of the sun. Unlike the light rays that we see, neutrinos come to us directly from the sun's interior. After they are produced, only a few of them interact with any of the sun's constituent atoms. The great majority stream unimpeded into space.

THE SOLAR NEUTRINO PROBLEM

In 1968, Raymond Davis, Jr., a physicist from the Brookhaven National Laboratory, began an experiment in the Homestake Gold Mine in Lead, South Dakota. He constructed a 100,000-gallon tank at a depth of 4,850 feet—nearly a mile beneath the surface of the earth—and filled it with ordinary cleaning fluid. Then he set to work to see how many solar neutrinos he could detect in the tank.

The cleaning fluid was a liquid called perchloroethylene, a compound made up of the elements carbon and chlorine; in each molecule there are two carbon and four chlorine atoms. Davis chose this substance because, on rare occasions, a neutrino will interact with chlorine to produce an atom of argon 37. If Davis could measure the quantity of argon that was produced, he would be able to calculate how many neutrinos had entered the tank. He expected that the many trillions of neutrinos that entered his apparatus would produce argon atoms at about the rate of two per day.

Now it so happens that argon 37 is radioactive. This fact is quite important. After all, hunting for one or two atoms of argon gas in a 100,000-gallon tank would be a hopeless task; finding the proverbial needle in a haystack would be much, much easier. But fortunately an atom of argon 37 will emit an electron when it decays, and individual electrons can be detected quite easily. All that Davis had to do was to set up an electronic apparatus that would count them. Davis was confident that his experiment would detect the neutrinos that were presumably coming from the sun, and only such neutrinos. He set up the apparatus far below the surface of the earth in order to eliminate the contaminating effects of reactions induced by cosmic rays. Neutrinos could easily penetrate to the depth at which the experiment was set up, but the particles that constituted cosmic "radiation" would not. Thus if Davis saw what looked like a neutrino-endured reaction, he could be reasonably sure that it had been induced by a neutrino from the sun.

Davis was confident of this because, although the universe is almost certainly full of cosmic neutrinos that bombard the earth from all directions, these cosmic neutrinos (which are relics of the big bang) do not possess enough energy to convert chlorine into argon. Finally, Davis checked his equipment to verify that he could indeed find any argon 37 atoms that were produced; and he found they could be detected with an efficiency of about 90 percent.

Then the experiment was run. The results were not as expected. Davis found that the number of neutrinos coming from the sun seemed to be only about 10 percent of the quantity that theory predicted.

CONSTERNATION IN THE COMMUNITY OF ASTROPHYSICISTS

This result caused consternation among astrophysicists, who immediately set to work rechecking their theoretical calculations. They soon determined that small changes in the assumptions about conditions (such as temperature and pressure) in the center of the sun could cause large variations in the rate at which neutrinos were created. So the theorists altered some of their figures and came up with a new estimate for the neutrino production rate. But even with these revised figures, the estimate of neutrinos that should be observed was quite a bit higher than Davis's results. "Manipulating" figures in this way may sound like a somewhat unscientific procedure to the lay reader. So I should probably point out that this was a legitimate process. New experimental results often make it necessary for scientists to refine their theoretical assumptions, just as new theoretical discoveries often suggest experiments that should be performed. It is simply an example of one of the ways in which theory and experiment interact.

In the meantime, Davis continued his experiments and obtained more accurate results that produced an even lower estimate of neutrino production than the figures he had originally announced. Work continued in both the experimental and theoretical arenas. In the end, after Davis had labored as hard as he could to find all the neutrinos that were there, and after the theoretical astrophysicists had worked as hard as they could to explain the lower number, there was still a huge gap between experiment and theory. The number of neutrinos Davis counted was no more than a third of the number that should have been observed according to theory.

Since 1968, the Davis experiment has been repeated a number of different times in several different ways. Neutrino detectors have been set up in a mine near Kamioka, Japan; in Russia's North Caucasus region; and in the Gran Sasso tunnel below the Italian Apennines. Each of these experiments yielded different specific results because each made use of different kinds of neutrino detectors, which see neutrinos in different energy ranges. There is no reason why the gap between theory and experiment would have to be the same for low-energy as for higher-energy neutrinos, after all. The important thing is that every one of the experiments produced results which fell below theoretical predictions.

Scientists have proposed a number of explanations for the low neutrino count, but as yet no one knows which one is most plausible. Two possibilities seem especially likely. Either there is something wrong with the theories about the conditions that exist within the interiors of the stars and fewer neutrinos are created than scientists supposed, or some of the neutrinos have been undergoing transformations between the time they leave the sun and the time they arrive at a detector under the surface of the earth. More precisely, it is possible that one kind of neutrino may be changing into another.

Three Kinds of Neutrinos

When neutrinos were first detected in 1956, scientists assumed that they were all of one kind. Today, they know that there are three distinct neutrino varieties. The most common type—those produced in the sun—are called electron neutrinos because they participate in reactions that also involve electrons. The other two kinds are seen in reactions involving particles known as the *muon* and the *tauon*. These two particles, which take their names from the Greek letters *mu* and *tau*, have properties similar to those of the electron, but they are much heavier. In fact, it would not be too inaccurate to describe them as "heavy electrons." Although the muon and tauon (and their associated neutrinos) are infrequently observed in nature, both particles can easily be produced in the laboratory.

No one is really sure whether an electron neutrino can spontaneously change into a neutrino of the muon or tauon variety or not. However, there are theoretical reasons for believing that such transformations might be possible. If they are, this could not only provide a solution to the solar neutrino problem but could also explain the discrepancy between the different experimental results. After all, it is logical that electron neutrinos in certain energy ranges might change more readily than others. If such changes took place, the experiments would simply not "see" the muon and tauon neutrinos. The muon neutrino is relatively difficult to detect, and the tauon neutrino has not yet been observed at all; its existence is purely theoretical.

On the other hand, there is no experimental evidence that anything like this does actually take place. Even the theoretical calculations don't suggest a definite answer. All that currently accepted theory does is to suggest that such neutrino "oscillations" could happen if the neutrino has a small mass. And it has not yet been determined whether the neutrino mass is very small or whether it is precisely zero. It is very difficult to measure the properties of an elusive particle like the neutrino which interacts with other particles of matter so rarely.

Consequently, it is necessary to remain open to the possibility that there might be something wrong with astrophysicists' ideas about conditions in the interiors of stars. Though it is very probable that accepted theories are correct in their broad outlines, there may be things going on in stellar cores of which scientists are unaware. If so, then it is obviously not possible to have complete confidence in estimates of the ages of old stars.

Thus pumping 100,000 gallons of cleaning fluid into a tank situated at a depth of nearly one mile has turned out to have a direct bearing on attempts to estimate the age of the universe. If the solar neutrino problem is eventually cleared up, it may be possible to say with confidence that globular cluster stars are indeed as old as they seem to be. But, until this happens, it is necessary to entertain the possibility that the chain of reasoning that led to a figure of 15 billion years for these stars may contain undiscovered flaws.

However, if there are unanswered questions about the methods used to determine the ages of stars, it does not necessarily follow that it is impossible to obtain an accurate estimate of the age of the universe. As we shall see, there are a number of different ways to approach this problem and a number of different kinds of evidence to be examined. If the various different results can be reconciled, it might be possible to obtain an estimate that is accurate indeed.

For example, why not measure the age of matter itself?

THE AGE OF THE ELEMENTS

Uranium 238 is a radioactive isotope that has a half-life of 4.5 billion years. What this means is that, if one were to take a kilogram of this substance, only half a kilogram would remain after that period of time; the other half would have decayed to an isotope of lead. Then, after a second 4.5 billion years had passed, half of what was left would have decayed, and only a quarter of a kilogram would remain.

The decay of uranium is a rather complicated process that takes place in fourteen steps. The first thing that happens is that a uranium 238 nucleus emits an alpha particle (an alpha particle is the same thing as a helium nucleus; it consists of two protons and two neutrons), decaying into thorium 234. The thorium then emits an electron and becomes protactinium 234. Next, another electron is emitted and one has uranium again, but this time it is a lighter isotope, uranium 234, The next three steps produce thorium 230, radium 226, and radon 222. Then, after various further decays, lead 206 is finally created. Since this isotope of lead is not radioactive, the series ends at this point.

Knowledge of the half-life of this process makes it possible to determine the age of rocks by measuring the amounts of uranium and lead that they contain. Since chemical and physical processes would have caused any uranium that was originally present to crystallize out as a relatively pure substance when the rock was formed, any lead that is found to be mixed in with the uranium must have been created by radioactive decay.

When this method is used to find the ages of objects in the solar system, it gives remarkably consistent results. Dating of rocks from the moon and dating of the oldest meteorites found on the surface of the earth both give ages of about 4.6 billion years. The oldest terrestrial rocks that have been found have proved to be 3.9 billion years old. And this is just about what one would expect. The earth, after all, was originally in a molten state, and it would have had to have existed for some time before it began to form a solid crust. The moon, on the other hand, would have cooled more quickly because of its relatively small size.

It is therefore possible to conclude that our solar system was formed 4.6 or 4.7 billion years ago. Naturally this tells us very little about the age of the universe, since there is no way of telling how much time passed between the big bang and the creation of our sun. However, it does give us confidence that radioactive dating can be used to obtain reasonably accurate ages. This success has encouraged scientists to go a step further and see if they can use radioactive dating to determine the ages of the elements themselves. In other words, if one can use uranium-lead dating to determine when the uranium was first incorporated into a rock, could there not be some kind of procedure that would tell us at about what date the uranium itself was created?

Such methods exist. Although they are more complicated than the one I have described above, the principle is the same. One can determine the age of the elements by comparing the relative abundance of two different radioactive substances and then comparing these to a primordial abundance that is calculated theoretically. For example, suppose there are reasons for thinking that radioactive substance A was originally created in quantities that were twice as great as radioactive substance B. If so, the ratio of the two substances will not continue to be 2 to 1 because they will have different half-lives: One of them will decay more rapidly than the other. Therefore, if we find that they exist in a ratio of, say, 4 to 1 now, it is possible to calculate how much time has passed since the ratio was 2 to 1.

As one would expect, the uncertainties associated with this method are greater than those encountered in the dating of rocks, and the results are consequently less precise. However, one can use radioactive dating to estimate an age of about 11 billion years for the radioactive elements. Since there is an uncertainty of plus or minus 1.5 billion years, one can only conclude that these elements are between 9.5 billion and 12.5 billion years old.

This does not immediately provide us with an estimate for the age of the universe, however, because (as we saw in Chapter 1) the heavy elements were not created in the big bang. But at least we can come up with a lower boundary on the age of the universe. If the radioactive elements are at least 9.5 billion years old, then the universe has to be older than that.

If the heavy elements were not created in the big bang, when were they created? Our sun is a second-generation star. Like the planets that surround it, it incorporates a great deal of cosmic debris that was spread through space when some first-generation stars exploded as supernovas. After galaxies had formed, gravity created stars by further condensing the primordial hydrogen and helium gas. These stars contained little but a mixture of these two gases. At this time, only hydrogen, helium and small amounts of certain other light elements that had been manufactured in the big bang existed. There were no heavier elements because the conditions for making them did not yet exist.

Many of the stars that were created during this era still exist. In fact, the chemical composition of stars found in globular clusters that they are such first-generation stars. But of course some of the stars that were created at this time ceased to exist long ago. The more massive stars consumed their nuclear fuel very quickly; some of them were destined to have life spans, not of billions, but only of hundreds of millions of years.

As these massive stars went through their death throes, the heavier elements (including such common ones as carbon, oxygen, and iron as well as such radioactive isotopes as uranium 238) were manufactured in the nuclear reactions taking place in the stars' cores. Then, when these stars exploded as supernovas, these elements were spread through space. After some unknown period of time, they were incorporated into second-generation stars like our sun, and into the planets that were created around them.

If one assumes that it took 2 billion years or so before the first stars were created, one comes up with a minimum age of 12 billion years or so for the universe. It is not possible to say how confident one can be about this figure. It could be that some of the assumptions that yielded an age of between 9.5 and 12.5 billion years for the radioactive elements are questionable. However, this age does seem to be consistent with an age of 9 billion years for the oldest white dwarf stars. In fact, if the white dwarfs are second-generation stars, this is just about what one would expect.

THE HUBBLE CONSTANT

A neat little picture seems to have developed. There appear to be old stars that are 15 or 16 billion years old and younger ones that are up to 9 billion years old. Since the heavy elements apparently were created about 11 billion years ago, they could easily have been created in the more massive of the older stars and then incorporated into the younger ones. It would appear that the evidence points to a universe with an age of at least 17 or 18 billion years, adding in 2 billion years for the formation time of the oldest stars.

Unfortunately, matters are not that simple. Other evidence indicates that the universe may be no more than about 12 billion years old, and probably somewhat less. In fact, according to some astronomers, the true age of the universe could be as little as 8 billion years.

As I pointed out in the previous chapter, there is a considerable amount of uncertainty about intergalactic distance scales, and hence about the rate at which the universe is expanding. Because the rate of expansion is one of the measures used to estimate the age of the universe, the uncertainty about distance scales leads, in turn, to wide variations in the estimates of the time that has elapsed since the big bang. However, the very fact that such uncertainties exist has induced astronomers to refine their methods in an attempt to obtain more accurate results. During the five or ten years that preceded the writing of this book, astronomers seemed to be reaching a consensus that the universe was relatively young. The premise on which they worked was that one could obtain an accurate estimate of the universe's age by measuring its rate of expansion. If one knew how rapidly it was growing, it should not be difficult to extrapolate back to time zero and to determine when the expansion had started. Admittedly, the results obtained in this way were at variance with determinations of the ages of stars. However, measurement of the expansion itself was presumably a more direct, and therefore more reliable, method. Using it required that fewer assumptions be made.

A few weeks before I began writing, a group of astronomers announced results that were totally at variance with the consensus figures. They argued that data obtained by means of the Hubble space telescope indicated that the universe could be much older than the majority of astronomers believed it to be. Naturally this led to a great deal of controversy, which is still raging as I write this. Before I explain what the controversy is about, however, it will be necessary to define a few terms and to explain something about the scientific background underlying the controversy.

When Edwin Hubble announced his discovery that the universe was expanding in 1929, he backed up his claim by pointing out that the speed of recession of a galaxy was proportional to its distance from the earth. If galaxy A was twice as far away as galaxy B, it proved to be receding at a speed that was twice as great. And if it was, say, six times the distance, then it would be found to be moving six times as fast. As Hubble pointed out, this was exactly the kind of behavior one would expect to observe in an expanding universe.

The fact that distance and speed of recession are proportional to one another makes it possible for astronomers to define a quantity called the Hubble constant, a number which is obtained by dividing the speed of recession of a galaxy by its distance from the earth. Since, as far as astronomers can tell, the Hubble constant is the same for very distant galaxies as for ones that are relatively close to us, it is a number that describes properties of the universe itself as well as the objects (galaxies) in it.

The Hubble constant is normally expressed in units of kilometers per second (velocity) divided by megaparsecs (distance). A megaparsec is equal to about 3 million light years. It is a convenient unit for describing the distance of galaxies that may lie hundreds of millions or billions of light years away. It will not be necessary to remember anything about kilometers per second or megaparsecs to follow the ensuing discussion, however. The only point that needs to be remembered is that, if one knows the Hubble constant, then the age of the universe can be estimated.

In recent years, most estimates of the Hubble constant have ranged between 50 and 100 (that's 50 to 100 kilometers per second per megaparsec, but we have already agreed to drop the units). A Hubble constant of 50 would correspond to a universe that was 15 or 20 billion years old, and a constant of 100 would correspond to an age between 7 and 10 billion years. A large Hubble constant (100, say) corresponds to a relatively young age for this reason: If we find that the universe is expanding very rapidly, then we know that there hasn't been enough time for it to slow down very much; on the other hand, if we observe a relatively slow expansion, then we can conclude that a lot of time has passed since the big bang.

At this point, a skeptical reader might ask why knowing the Hubble constant gives only a range of ages, not an exact figure. The reason is a relatively simple one. Since the rate at which the expansion is slowing down depends on the quantity of matter that the universe contains, one must know both the Hubble constant *and* the matter density to compute an age.

Calculations indicate that if the density of matter is close to the critical density,* then the universe is younger than it would be if the density were very low. For example, I pointed out above that a Hubble constant of 100 implies an age between 7 and 10 billion years. If I wanted to be more precise, I could say that one would be able to conclude that the universe was 7 billion years old if the matter density was equal to the critical value, and that it was 10 billion years old if the density was found to be very low.

So, even though there are a number of uncertainties involved, it is important to obtain an accurate estimate of the Hubble constant. If astronomers were able to determine that the age of the universe was somewhere in the 7- to 10-billion-year range or, alternatively, that it was between 15 and 20 billion years, they would have accomplished a great deal. The latter figure, for example, would tend to confirm calculations that have led to estimates of 15 or 16 billion years for the ages of old stars.

Unfortunately, the matter has not been settled. By the early 1990s, astronomers seemed to be coming to the conclusion that the Hubble constant was probably somewhere around 85, and that the age of the universe was therefore somewhere between 8 and 12 billion years. To be sure, some held out for a Hubble constant that was lower or higher. However, the number 85 was frequently touted as a good compromise figure.

An age of 8 to 12 billion years was not consistent with estimates of 15 or 16 billion for the ages of old stars, however. To make matters worse (and I will discuss this matter in detail in the next

* Recall that the critical density defines the borderline between an open and closed universe, and that the retarding forces of gravity are greater in the latter.

chapter), there were theoretical reasons for believing that the matter density of the universe was close to the critical value (the one that determines the borderline between an open or closed universe). If so, the amount of time that had elapsed since the big bang was much closer to 8 billion years than it was to 12 billion. Obviously, something was wrong. The idea of 16-billion-year-old stars in an 8-billion-year-old universe was ridiculous. To be sure, one could fudge a bit and use a "ballpark" figure of 10 billion rather than 8. But that was hardly much better.

TYPE IA SUPERNOVAS

Then, early in 1992, an astronomer named Allan Sandage published some new results. Sandage was an astronomer of the older generation. Born in 1926, he had once worked as Hubble's assistant. He had spent years attempting to measure the expansion of the universe, and had developed many of the methods used by astronomers attempting to measure intergalactic distances.

Sandage had long maintained that the true value of the Hubble constant was about 50 (he had even been known to joke that God had told him that this was the correct number) and that the universe was consequently much older than the majority of his colleagues believed. Now, in 1992, he believed that he had finally found proof of this contention. Together with his colleagues F. Duccio Macchetto, Nino Panagia, and Abhijit Saha of the Space Telescope Institute in Baltimore and Gustav Tammann of the University of Basel in Switzerland, he presented evidence that seemed to indicate that the Hubble constant was only around 45. He based this conclusion on Cepheid variable stars.

Cepheid variable stars are among the more accurate of the intergalactic distance indicators. As I pointed out in Chapter 2, the intrinsic brightness of these stars is known. Consequently, when astronomers discover a Cepheid in a distant galaxy, they can easily compute the galaxy's distance from the earth. The dimmer the Cepheid variable appears to be, the farther away it is.

Before Sandage's discovery, this method could only be used at distances up to about 10 million light years. Beyond that distance, the Cepheids were too dim to be seen. After the Hubble space telescope was placed in orbit, however, Sandage and his collaborators were able to use this telescope to make out Cepheids in a galaxy called IC 4182 that was located 16 million light years from the earth.

Extending a distance scale from 10 to 16 million light years doesn't sound like a major accomplishment, and it wouldn't have been, except for one important fact. It so happened that astronomers had observed a type Ia supernova in ICC 4182 in the year 1937. As we shall see, this observation combined with Sandage's yields some real information.

Type Ia supernovas occur only in binary star systems in which one of the two stars is a white dwarf and the other is a red giant. If the two stars are close enough to one another, gravity will pull material from the red giant star and cause it to fall onto the surface of the dwarf. This will cause the latter star to gradually grow in size. At first, nothing much happens. But then, when the white dwarf attains a mass approximately equal to 1.4 times the mass of our sun, increasing temperatures and pressures cause it to become unstable and is suddenly transformed into a massive thermonuclear bomb. When the mass of the white dwarf reaches that critical figure, it explodes as a type Ia supernova.

If all type Ia supernovas have the same mass, then they should all be equally bright. So if one compares the apparent brightness of type Ia supernovas in two galaxies, the relative distance of the two galaxies can easily be determined.

But it is not enough to know that one galaxy is two or four or ten times as far away as the other. If the Hubble constant is to be accurately determined, one must also know absolute distances. And this is where the Cepheid variables come in. By extending the range of the Cepheid variable method, Sandage and his colleagues were able to calculate the distance of a galaxy in which a type Ia supernova had occurred. This, in turn, allowed them to calibrate a distance scale that gave the distances to many other galaxies. When they did this, they found that the Hubble constant was somewhere between 30 and 60, with 45 the most probable value. They hadn't found the constant exactly; after all, even the best astronomical distance measurements contain sources of error. But they had found a value that was at odds with what was rapidly becoming an accepted figure.

The results were somewhat tentative. Even Macchetto, one of the members of the team, admitted that he was "uncomfortable" about basing such a result on observations of just one supernova.

He pointed out, however, that the method had the potential to be extremely accurate. After all, there were other Cepheids and other type Ia supernovas. More observations could be done in the future, and, as they were, more precise figures could be obtained.

However, the team had come up with a range of values for the Hubble constant that was far from the accepted value of 85. And so controversy flared to an even more intense level. For one thing, some astronomers did not agree with the measurements of the distance to IC 4182 that the Sandage-Macchetto team had obtained. For example, a group of astronomers led by Michael Pierce of Kitt Peak National Observatory had used red giant stars to measure the galaxy's distance and had obtained a result of 8.5 million light years, not 16 million. If it was correct, Pierce's result would give a Hubble constant of 86.

An interesting sidelight on the controversy is provided by an article entitled "The Hubble Constant" that appeared in the journal *Science* just a short time before the Sandage-Macchetto result was announced. In this article, astronomer John P. Huchra of the Harvard-Smithsonian Center for Astrophysics included a table of "recent measurements" of the Hubble constant. Some thirty different results were listed. The smallest was a value of 42 (with an uncertainty of plus or minus 11). The largest value found for the Hubble constant was 105 (again with an uncertainty of plus or minus 11).

It appears that, if one takes all the various different findings into account and figures in all the uncertainties, it is only possible to conclude that the Hubble constant is somewhere between 30 and 116. If one uses these figures to calculate the age of the universe, one reaches the conclusion that it must be at least 6 billion years old, and is probably no older than 30 billion.

In practice, things aren't that bad, however. Some methods are more accurate than others, and the more extreme values for the Hubble constant can almost certainly be discarded. If one were to say that the true value is probably between 50 and 90, one would probably not be contradicted by any future results.

On the other hand, it seems likely that years will pass before the Hubble constant is known to any great degree of accuracy. And until it is, scientists will know neither how fast the universe is expanding nor how old it is. Even though astronomers have been trying to find a way to accurately measure the Hubble constant for more than sixty years, they are still far from achieving success.

4

How Will the Universe End?

When Einstein first became famous, roughly around 1920, the public tended to think of him as a mathematician rather than as a physicist. He was viewed as a mysterious personage who was somehow able to unravel the secrets of the universe by writing down arcane mathematical equations.

There was some justification for this point of view. Theoretical physics is highly mathematical, and it has become so because if theoretical premises can be put in mathematical form, then mathematical methods can be used to draw conclusions. Mathematics, after all, is a variety of logical reasoning. To be sure, all that mathematics will do is link assumptions and conclusions together; it will not tell scientists whether their assumptions are correct. However, the correctness of a theory can be tested by performing experiments.

One of the reasons that mathematics is so useful is that it sometimes leads to unexpected conclusions. Scientists are often able to discover phenomena that they would never have thought to look for if they had depended on verbal reasoning alone. A good example is provided by the general theory of relativity, which tells us that the existence of pressure in a medium—defined as an innate tendency to expand—will increase the gravitational force operating in the system. In other words, if a system has a tendency to expand outward, it will experience attractive gravitational forces that will induce it to contract.

At first, this result appears counterintuitive. After all, when one blows up a balloon, the air pressure that is created causes it to expand, just as one would expect. Forcing air into a balloon never causes it to shrink. However, after all, a balloon contains a relatively small amount of matter, and the gravitational forces that air molecules exert on one another are so small as to be totally insignificant. There are good reasons for not trying to apply a theory of gravity like general relativity to an object that weighs only a few grams.

On the other hand, it is quite reasonable to apply the general theory of relativity to the behavior of a massive star. In such a case, gravitational forces are very strong because there is so much mass. The existence of pressure will then cause the inward force to become greater. For example, when a large star is dying and undergoing gravitational collapse, the existence of a great deal of pressure will not cause the star to expand like a balloon. In fact, it would not even slow the collapse. On the contrary, the existence of outward pressure will create an additional inward force that will make the collapse even more rapid.

As I mentioned in Chapter 2, the gravitational effects of pressure play a role in the formation of black holes. The existence of pressure creates such significant additional inward forces that nothing—not even light—can escape. Without these forces, the dying star might shrink into a neutron star (a compact body in which individual protons and electrons no longer exist because they have been "squeezed" together to form neutrons), and black holes would be formed only rarely or not at all.

NEGATIVE PRESSURE

If the existence of outward pressure (the innate tendency to expand) can produce gravitational forces that are directed inward, then there is every reason to think that negative pressure should produce forces that are directed outward. Or at least this should happen under certain circumstances.

The concept of negative pressure is not so arcane as one might think. It is nothing more than an innate tendency to contract. For example, in an automobile engine, positive pressure is created when a gasoline-air mixture in the cylinders is ignited. The pressure forces the piston outward, creating the power that causes the car to run. On the other hand, if the gasoline-air mixture were not ignited but instead suddenly cooled, the vapor in the cylinder would tend to contract, and the piston would be drawn inward. It is doubtful that one could run an automobile by this method, but it would be an example of negative pressure.

Another example that is sometimes given is that of a solid rubber ball that has been pulled outward in all directions (it isn't specified how this feat is accomplished, but perhaps it could be done somehow). The ball would then have a tendency to contract, and one could say that negative pressure existed within it.

Obviously, it would be as silly to attempt to apply the general theory of relativity to a small rubber ball, or to an automobile cylinder, as it would be to try to use it to describe the behavior of a balloon. Nor does the concept seem especially relevant to the behavior and evolution of stars. Stars exhibit positive pressure. In fact, under ordinary conditions, it is this pressure that prevents them from collapsing (positive pressure may create an inward gravitational force, but this force is not sufficient to cause contraction under ordinary conditions). So at first glance, the idea of negative pressure appears to be one of those concepts that is perfectly logical but not very useful.

THE INFLATIONARY UNIVERSE THEORY

This would indeed be the case were it not for the fact that negative pressure may have played a significant role in the evolution of our universe. In fact, there is good reason to believe that the negative pressure that existed when the universe was 10^{-35} seconds* old produced features that astronomers observe today.†

* 10^{-35} is the number represented by the numeral "1" followed by thirty-five zeros. 10^{-35} is the number 1 divided by 10^{35} and is thus an extremely small number. It could also be

The significance of negative pressure was first noted by the MIT physicist Alan Guth (the name rhymes with "tooth"). In 1980, Guth pointed out that certain theories of subatomic particles, known as grand unified theories, or GUTs (more on these shortly), seemed to imply that there existed a great deal of negative pressure in the early universe. If so, Guth noted, the universe would have undergone a period of rapid "inflationary expansion." (See Figure 6.) During this period, the universe might have increased in size by a

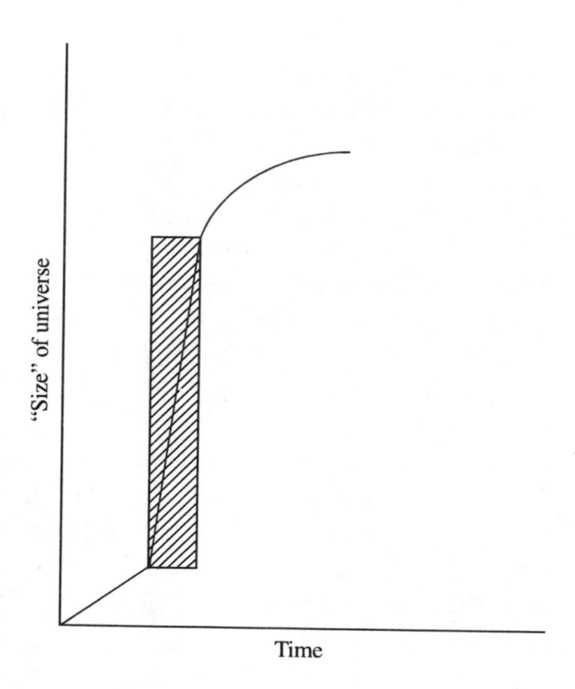

Figure 6 *The shaded region represents an inflationary expansion. For a brief period of time, the universe expands at a greatly increased rate. The diagram is not drawn to scale; if it were, the shaded region would have to be at least billions of light years high. The onset of the inflationary expansion of the universe was so dramatic an event that it cannot be accurately represented in any drawing.*

---

expressed as "one hundred billionth of a trillionth of a second."

† I hope that the reader will not be disconcerted to find me talking about the first 10^{-35} seconds of time in a chapter entitled "How Will the Universe End?" A discussion of the fate of the universe will come in good time. However, it is necessary to first say a few things about the manner in which the universe has evolved. In order to do this, it is necessary to say some things about the conditions that existed at the beginning.

factor of 10^{50} or more, even though the inflationary expansion most likely lasted for no more than 10^{-32} seconds.*

Guth's inflationary universe theory occupies a strange place among the theories of contemporary science. There seems to be little or no empirical evidence that could be used to support it. To make matters worse, the GUTs upon which Guth's hypothesis is based are themselves quite speculative. No one knows which of the GUTs (there are a number of competing theories) is most likely to be correct, if indeed any of them are (none has so far proved to be wholly successful). And yet the inflationary universe idea has gained wide acceptance among scientists because certain features of the present-day universe seem impossible to explain if one does not assume that an inflationary expansion once took place.

Grand Unified Theories

Before I discuss the relevance of the inflationary universe theory to currently accepted ideas in the field of cosmology, we should backtrack a bit to discuss the GUTs and the role played by negative pressure a little more fully. As we do this, it will become apparent that, even though none of the GUTs has been experimentally confirmed, these theories are entirely plausible. Furthermore, they seem able to explain a feature of the universe that would otherwise be hard to understand: The fact that the universe contains little or no antimatter.

The GUTs are theories which attempt to give a unified explanation of three of the four forces of nature. These four forces are gravity, electromagnetism, and the strong and weak nuclear forces. The strong force is the force that binds neutrons and protons together in the nuclei of atoms, and the weak force is responsible for certain kinds of radioactive decay. The GUTs represent an attempt to combine descriptions of all the forces except gravity in a single theory.

Ideally, physicists would like to find a unified explanation of all four forces. Indeed, there are theories, called superstring theories, which attempt to do just this. However, superstring theories are even more speculative than the GUTs. And many theoretical

* Note that 10^{-32} (1 divided by 10^{32}) is a larger number than 10^{-35}. Thus 10^{-32} seconds is a longer time than 10^{-35} seconds, a thousand times longer, as a matter of fact.

physicists suspect that decades may pass before any superstring theory can be shown to be correct. In fact, superstring theory is sometimes described as twenty-first-century physics discovered by accident during the twentieth century.

Discussing superstring theory in detail would make it necessary to stray too far from the main themes of this book, so I will say no more about the matter at this point. However, a few comments on physicists' motivations for seeking unified explanations of the forces might be in order.

Obviously, it should be nice to have a single theory that showed that three or four different forces were really only different aspects of a single "superforce." That would make life somewhat less complicated, since one could then work with just one or two theories rather than with four different ones. But this is only part of the reason why scientists are attempting to find a unified explanation of the forces.

When a theory is found that provides a unified explanation for a number of seemingly disparate phenomena, new discoveries are likely to follow. For example, when the Scottish physicist James Clerk Maxwell showed, in 1873, that electricity and magnetism were only different aspects of a single force, he was able to deduce that electromagnetic waves should exist. After he did this scientists were not only able to conclude that light was a form of electromagnetic radiation, but they also realized that there should be other—as yet undiscovered—kinds of electromagnetic waves. Thus Maxwell's discovery led directly to the invention of the radio.

Similarly, when the American physicist Steven Weinberg and the Pakistani physicist Abdus Salam proposed, in 1967, that descriptions of the electromagnetic and weak forces could be combined within a single theory, they were able to deduce the existence of certain subnuclear particles that physicists had not yet observed. All three of the particles that they predicted, called W^+, W^-, and Z^0 (the symbols indicate that the W particle can have either a positive or a negative charge, and the Z is electrically neutral), were discovered in 1983.

Although none of the GUTs has been confirmed by experiment, and although these theories seem to be plagued with certain theoretical difficulties as well, physicists still find the ideas on which they are based to be quite appealing, in part because they seem to represent the next logical step in the unification of the forces. They also seem capable of explaining why the universe should contain so little antimatter.

MATTER AND ANTIMATTER

For every particle that physicists have discovered, there is a corresponding antiparticle. There are protons and antiprotons, neutrons and antineutrons, and electrons and positrons (the positron should really be called an "antielectron"; however, it was given the name *positron* before present-day nomenclature came into existence). There are antimuons, antineutrinos, and anti-W and anti-Z particles as well.

Antiparticles can easily be created in the laboratory. According to Einstein's equation $E = mc^2$, it should be possible to create matter out of energy. Indeed, scientists accomplish this feat every day. But antiparticles never appear alone. Whenever mass is created, it always appears in the form of particle-antiparticle pairs. A proton will appear in the company of an antiproton, for example. An electron and a positron will pop into existence together, and so on.

This raises the question of why the universe should be made predominantly of matter, and why antimatter should be rare or nonexistent. Here, when I speak of "matter," I mean objects made up of the familiar protons, neutrons, and electrons. Antimatter would be a material in which atoms were made up of antiprotons, antineutrons, and positrons. It should theoretically exhibit properties similar to those of matter. There is no reason why there could not be antimatter stars, antimatter planets, and antimatter galaxies. For that matter, it is perfectly reasonable that there could be living organisms made of antimatter.

However, scientists have not observed significant quantities of antimatter in our universe. And they are reasonably certain that they would see it if it were there. When a particle and an antiparticle come into contact with one another, they mutually annihilate one another and disappear in a burst of energy. Thus if two objects came into contact with one another and one happened to be made of matter while the other was antimatter, the quantity of energy released would be enormous. For example, if an antimatter meteor entered the earth's atmosphere, all of its mass—and an equal quantity of matter—would be converted into energy. The result would be an explosion many times more powerful than the largest existing hydrogen bombs.

Meteors enter the earth's atmosphere all the time, and yet this never happens. Nor are such explosions observed on any of the other planets. Finally, the solar wind, which consists of subatomic particles that stream outward from the sun, does not participate in any matter-antimatter annihilations. Therefore, we can be reasonably certain that there is no antimatter within our solar system.

For that matter, as far as astronomers can tell, there are no significant quantities of antimatter anywhere in our galaxy. If there were, it could most likely be observed. For example, if there existed any interstellar gas that was made of antimatter, enormous quantities of energy would be released when this gas came into contact with gas clouds made of ordinary matter, or with ordinary matter stars. Similarly, if there were antimatter stars, then they would sometimes come into contact with ordinary matter gas.

Presumably, entire galaxies could be made of antimatter. This possibility is not very likely, however. Galaxies sometimes collide with one another, after all. And astronomers have never observed the energy release that would result if an antimatter galaxy and one made of matter came into contact with one another.

The fact that antimatter is not observed creates a very troubling problem. In fact, it raises the question why there should be any matter at all. During the early stages of the evolution of the universe, matter and antimatter would have been created out of energy, and they should have been created in equal amounts (as we have seen, a particle and an antiparticle are always created together). Why didn't all the matter and antimatter eventually annihilate each other, leaving a universe that contained nothing but energy?

We might assume that the universe began with an excess of matter over antimatter. This would certainly explain why matter should be predominant now. But making this assumption doesn't really solve the problem; it simply replaces one question with another. If we make this assumption, we are faced with the task of explaining exactly why there should have been more matter at the beginning.

GUTS TO THE RESCUE

The GUTs suggest a possible solution to this problem. They imply that when the universe was something of the order of 10^{-32} seconds old, there existed certain massive particles, known as X particles, which had the unique property that they decayed into particles *more often* than they decayed into antiparticles.

The scenario suggested by the GUTs goes something like this. When the universe was an infinitesimal fraction of a second old, large numbers of particles and antiparticles were created. Since many of these were created when X particles decayed, the numbers of particles and antiparticles were not equal. After all the X particles disappeared, there may have been something like a billion and one particles of matter for every billion particles of antimatter. The particles and antiparticles began annihilating one another (remember that the universe was very compressed at this point and that collisions would have been frequent); only the matter particles remained.

No one knows if something like this really happened. The X particle has not been discovered in the laboratory. Nor is it likely to be discovered during our lifetimes. If it really existed, it would be so massive—10^{14} or 10^{15} times heavier than a proton—that only a particle accelerator the size of our galaxy could produce it. And it is not very likely that any government is going to appropriate funds for a project of that magnitude.

But at least the scenario seems plausible, and no other convincing explanations for the preponderance of matter over antimatter exist. This fact gives physicists confidence in the idea that the GUTs might have something important to tell us about conditions in the early universe.

THE INFLATIONARY EXPANSION

The GUTs are theories of elementary particles: What could they possibly have to tell about the evolution of the universe, its beginning and its end? If correct, they might tell quite a lot, because the GUTs imply that space itself can possess varying amounts of energy. Or, as scientists often prefer to say, the universe can exist in different energy states.

The lowest possible energy state is known as the true vacuum. I hear the reader protest, "I thought we were talking about the universe, and the universe can be defined to be everything. What does this have to do with a vacuum, which is nothing?"

The universe is said to be mostly "empty" space. But space, as we have seen, is never really empty: Particles and particles are constantly being created and destroyed. The "vacuum" is therefore something that is full of matter and energy, and it can have properties which are quite important for the evolution of the universe.

What we call the true vacuum is the most stable state, and is presumably the one that exists today. (Or at least physicists hope that the universe is in this state. If it isn't, then the planet earth and all the stars could spontaneously disintegrate if the universe suddenly underwent a transition to another energy level.) All systems that are observed in nature seek the lowest possible level. For example, water will flow downhill until it reaches sea level. A hot object will cool until it reaches the temperature of its surroundings. Gas that is under pressure will flow out of its container when we turn the release valve.

According to the GUTs, the universe may not have been in the lowest possible energy state when it was about 10^{-35} seconds old; it may have been in a state called the false vacuum instead. Here the word *false* should not be taken literally. It is simply a piece of terminology that indicates that this was not the lowest, most natural level.

If the universe was in a false vacuum state, it would have made a transition to the true vacuum. But this would not have happened in an instant. Most likely the process took something of the order of 10^{-32} seconds. Although, 10^{-32} seconds would be an "instant" to most of us, things happened on a different time scale during that period. Then, 10^{-32} seconds was a long time when compared with 10^{-35} seconds.

So we come back to Guth's inflationary universe theory, which hypothesizes that the universe increased in size by a factor of 10^{50} or more during these 10-32 seconds. According to general relativity, the existence of a false vacuum would have created negative pressure that would have caused space to expand at this unimaginably rapid rate. Indeed, for a brief period of time, an expansive force like the one described by Einstein's cosmological constant may actually have existed. In fact, it would be quite accurate to say that the inflationary expansion took place because there was a large cosmological constant* in the early universe.

Well, all good things come to an end, and when the transition to the true vacuum was completed, the negative pressure would have disappeared. After that, the universe would have continued to expand, but at a much more moderate rate. The universe would now be coasting on its momentum; there would no longer be any outward force.

The idea that such an enormous expansion could have taken place in so short a period of time probably sounds bizarre. However, to call something "bizarre" in the fields of physics and cosmology is not a criticism. Anyone who pursues these disciplines encounters strange ideas every day. And anyone who deals with matter as far removed from everyday experience as the origin of the universe or the nature of matter on the subatomic level soon gets used to dealing with ideas that seem contrary to "common sense."

The question one should ask, then, is not whether the ideas associated with the inflationary universe theory are strange, but whether they are plausible. And when this is done, one discovers that it is quite a reasonable theory.

For one thing, the inflationary universe theory seems to explain where all the matter and the energy in the universe came from. If the theory is correct, then the universe might have contained little or no matter initially. Everything that we observe today—including the material from which the stars and galaxies were formed—would have been created spontaneously during the period of inflationary expansion. In fact, it would not be inaccurate to say that matter and energy would have rushed in to fill the rapidly expanding space.

I suppose that a statement like that may make the process sound somewhat mysterious, so perhaps I should restate some of the basic ideas involved. As we have seen, matter and antimatter can be created out of "nothing" if there is sufficient energy available. This energy would certainly have been available in an

* Speaking of "negative pressure" and of the existence of a "cosmological constant" are two different ways of describing the same thing. Both represent an outward force.

inflationary expansion, where spacetime was subjected to forces that were enormous. So there is every reason to think that large quantities of particles and antiparticles came into existence at this time. Furthermore, if the ideas associated with the GUTs are correct, it is conceivable that only matter was left over when the particles and antiparticles annihilated one another. Thus it is perfectly conceivable that the inflationary expansion could have taken a tiny universe that was nearly empty of matter and fashioned it into one like the universe we see around us today.

A FLAT UNIVERSE

The fact that the universe is filled with matter and energy today hardly constitutes proof that there was once an inflationary expansion. However, when one examines certain other features of the present-day universe in detail, it is hard to understand how they could have come about if the inflationary expansion did not take place.

For example, observations indicate that the matter density of the universe is almost certainly somewhere between 0.1 and 2.0 times the critical value (which defines the borderline between an open and closed universe). This is not a wide range, as we shall see.* At first, there may seem to be nothing particularly striking about this fact. Why shouldn't it be somewhere within that range of values?

The closeness of the observed density to the critical value is striking because the ratio changes with time. For example, suppose that the matter density of the present-day universe is one-tenth the critical density. If so, then there is not sufficient matter in the universe to halt the recession of the galaxies, and the expansion will go inexorably on. If it does go on, galaxies will become farther and farther removed from one another as time passes, and the retarding force of gravity will become progressively weaker. Matter will become progressively more dispersed, and as it does the matter density will fall to one-hundredth of the critical

* Recall that the critical density defines the borderline between an open and a closed universe.

value, then to one-thousandth. As the eons pass, it will decline still further, until eventually the density will be only one-millionth of that required to produce a closed universe, and then one-billionth, and one-trillionth.

We can use this same process of calculation to view the process in reverse and look back into the past. A straightforward calculation based on the estimate that the current density is 0.1 of the critical value shows that the matter density had to be very close indeed to the critical density when the universe was very young. Specifically, if the ratio is one-tenth now, then the matter density had to be equal to the critical density to an accuracy of one part in 10^{15} when the universe was one second old. So one is faced with the option of believing either that our universe is the product of some extraordinary coincidence, or that there was once a period of inflationary expansion.

And if there was an inflationary expansion, then the average curvature of space is very close to zero because an inflationary expansion would have eliminated any traces of spatial curvature that might initially have been present. Imagine blowing up a balloon. At first, the curvature of the balloon is very pronounced. If the balloon could be blown up to the size of the earth, however, one could not tell whether its surface was curved or not. After all, we are not aware of being on a curved surface unless we live right next to the ocean and can see ships disappearing over the horizon. And yet blowing a balloon up to the size of the earth would be an expansion that increased its directions only by an amount roughly equal to 10^8. This is many, many orders of magnitude smaller than the increase of 10^{50} or more that presumably took place during the inflationary era. (See Figure 7.)

OTHER PHENOMENA EXPLAINED BY THE INFLATIONARY UNIVERSE THEORY

Another phenomenon that only the inflationary universe theory seems capable of explaining is the fact that the universe looks pretty much the same in every direction: Or at least it does when one looks out to a distance of billions of light years. As far as astronomers can tell, all very large volumes of the universe contain

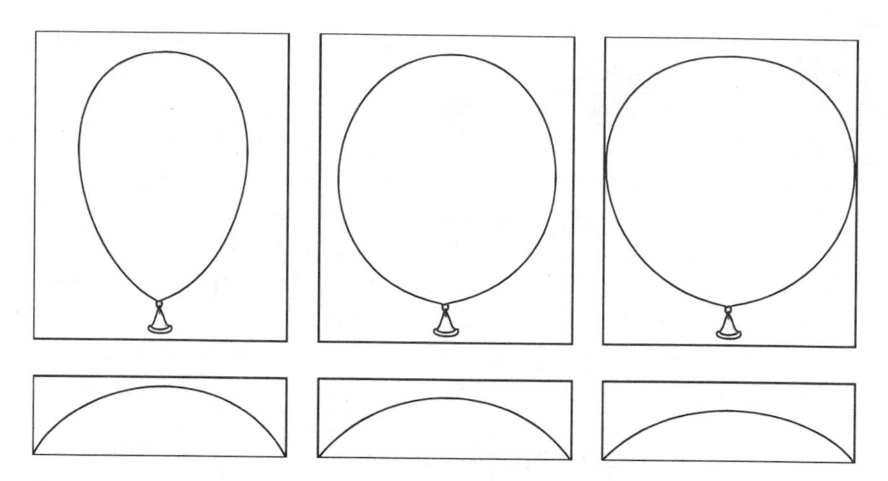

Figure 7 *The "flattening" of space as the universe expands can be represented by the blowing up of a balloon. Initially, the surface of the balloon is greatly curved. But as the balloon grows bigger and bigger, its surface becomes progressively flatter. In the last three panels, only a portion of the balloon is shown.*

approximately the same number of galaxies. To be sure, clusters and superclusters of galaxies are observed. But if one goes out far enough, the distribution of matter evens out.

One might compare the distribution of matter in the universe to that of sand on a beach. To an ant, the surface of the beach would appear uneven indeed. Individual grains of sand would take on the appearance of boulders. But a human being who looked out over several kilometers of sand would not be aware of such irregularities; he or she would see only a level expanse.

When one looks at the cosmic microwave background (see Chapter 1)—and doing this is equivalent both to looking very far back in time and looking very far out into space—the universe appears even more uniform. In fact, it is so uniform that it wasn't until the COBE results were obtained in 1992 that scientists found variations in the background at all, even though they had been looking for them for years. And then, as I noted in Chapter 1, the unevenness turned out to be only of the magnitude of one part in a hundred thousand.

The uniformity would be perplexing if there had been no inflationary expansion. The regions on the opposite sides of the universe could never have been in contact with one another had there been no expansion, and if they were never in contact with one another, it is hard to imagine how there could have been any "smoothing out" process that could have made them look so much alike.

Let us assume, for the sake of argument, that the universe is 15 billion years old. Suppose next that astronomers happen to view regions of space that are each 12 billion light years away on opposite sides of the earth. Since 12 + 12 is 24, these regions must be 24 billion light years apart. But if they are, they can never have been in contact with one another. According to the special theory of relativity, no signal or causal influence can travel faster than the speed of light, and as far as that took us was 12 billion. Therefore, there can have been no force—such as gravity—that could have acted on both regions and have caused such a smoothing out—unless one assumes inflationary expansion.

Admittedly these regions were once closer together. However, one still reaches the conclusion that they could not have had any influence on one another. At one time, for example, they may have been only 2.5 billion light years apart. But if the universe was only 1.25 billion years old at that point, each still would have been beyond the other's cosmic horizon. An observer in one would not have been able to see the other because there would not have been enough time for light signals to travel that far.

In an inflationary universe, on the other hand, there is no puzzle about why the two regions look the same. The inflationary universe theory implies that such regions would have been very close to one another before the period of inflationary expansion began. If they were, they could very well have been subject to the same "smoothing out" effect. In fact, the smoothing process could have been the work of the inflationary expansion itself.

Finally, the inflationary universe theory solves a problem that was once thought to be a good reason for being skeptical about the GUTs. All of the GUTs predict that magnetic monopoles should exist. A magnetic monopole is a particle which would have the appearance of an isolated north or south magnetic pole and which would have a mass about 1016 times greater than that of a proton. If magnetic monopoles exist, then north and south poles would not always appear together, as they do in magnets. According to the GUTs, magnetic monopoles should be about as abundant as protons.

However, this is not the case. If it were, the mass density of the universe would be about 1016 times the critical value. And this is impossible, because a universe containing that much mass would have collapsed in a big crunch long ago. The force of gravity would have been so great that the expansion would have continued only for a very short time before a period of rapid contraction set in.

Furthermore, magnetic monopoles have never been observed, or at least that is the consensus. In 1982, Stanford University physicist Blas Cabrera reported that he had detected a monopole or something that looked like one. However, this result has never been duplicated, either by Cabrera or by anyone else, and physicists tend to believe that it was probably spurious.

If one or another of the GUTs is correct, then where did all the monopoles go? The inflationary universe theory solves this problem too, by explaining that the inflationary expansion flung all the magnetic monopoles into the far reaches of space. The increase in the volume of space in the universe was so great that the density of monopoles was diluted to an absurdly low figure. According to calculations based on the theory, there should be no more than one or two magnetic monopoles in the entire observable universe.

BUT IS THE INFLATIONARY UNIVERSE THEORY TRUE?

Guth's theory is in some ways a maddening one. It appears to explain why the universe we live in possesses the characteristics that it does. And if the universe didn't have these characteristics, incidentally, we wouldn't be here to see it. A universe that was very far from critical density, for example, would most likely never produce life. Either it would collapse into a big crunch long before life had a chance to evolve, or its expansion would be so rapid that matter would never have a chance to condense into galaxies, stars, and planets.

On the other hand, there doesn't seem to be any hard empirical evidence to support this theory. It provides a very plausible scenario for the evolution of the universe, but it does not yield any testable predictions that can be confirmed by experiment or observation. It has gained such wide acceptance not because that it has been confirmed by experiment, but rather because no one has been has been able to think of any other reasonable way that a universe like ours could have evolved.

Though we may never solve this puzzle definitely, the inflationary theory might be confirmed indirectly. If one of the GUTs won out over its rivals, for example, and was confirmed by experimental evidence, then the idea that an inflationary expansion was inevitable would be on firmer ground. Even if this did not happen, the inflationary theory could gain some kind of firm theoretical support. Many scientists now believe that a number of different mechanisms could have brought about inflation. But the theories on which this idea is based also contain speculative elements.

As I write this, it is still possible to suggest that some alternative to the inflationary universe theory may yet be discovered. At present, one must admit that it appears that an inflationary expansion *probably* took place. But one must also bear in mind that widely accepted scientific ideas have frequently been overthrown in the past, and that similar events are likely to take place in the future.

THE AGE OF THE UNIVERSE AGAIN

Accepting the inflationary universe theory solves a number of problems, but it also seems to make at least one problem worse. This has to do with the age of the universe. The problem is a consequence of the fact that a universe in which an inflationary expansion once took place must have a matter density that is almost exactly equal to the critical value. But when we use this estimate of the matter density to calculate the age of the universe, we run into contradiction with other estimates of its age.

In order to understand how we can use measurements of the matter density to estimate the universe's age, consider that a universe that is right on the borderline is considerably younger than one with a density much less than the critical value. In fact, if the calculation is carried out in detail, one finds that it would be precisely two-thirds as old. It is not difficult to see why. A universe with a relatively small amount of matter is one in which the gravitational retarding forces are weak. Such a universe continues to expand at more or less the same rate over long periods of time. A critical density universe, on the other hand, contains more matter and must be slowing down more rapidly.

A universe which had an inflationary expansion, then, would have been expanding more rapidly in the past (note that I am not speaking of the inflationary period here, but of the expansion of the universe after inflation was over). But if the expansion was once more rapid than it is now, less time was needed to bring the universe to its present state.

The age of the universe depends not only on the matter density, but also on the size of the Hubble constant (the number that measures the rate at which the universe is expanding—see Chapter 3). Because there are a number of different estimates of the constant, it will be necessary to explore a number of different alternative estimates of the age of the universe. If the Hubble constant is equal to 85 (still the most commonly accepted figure) and the matter density of the universe is much less than the critical value, then one calculates that the universe is 12 billion years old. But if an inflationary expansion once took place and the density is very near critical value, then its age is only 8 billion years.

If, on the other hand, the true value of the Hubble constant is approximately 45, as Sandage and his collaborators claim, then a relatively empty universe would be somewhat more than 20 billion years old. And the age of a universe that had experienced an inflationary expansion would be about 15 billion years.

Thus we see that an inflated universe can be about 15 billion years old at best. And this is not consistent with estimates of 16 billion years for the ages of the oldest stars. One is forced to conclude either that there is something wrong with the inflationary universe theory or with astronomers' ideas about stellar ages.

On the other hand, 16-billion-year-old stars could easily exist in a noninflationary universe that could be 20 billion years old, or older. But of course, if the Hubble constant turned out to be closer to 85 than to 45, there would be a problem here too.

An inflationary universe can only be 8 to 15 billion years old. A noninflationary universe, which could have a matter density that was *either* very low or close to the critical value could be anywhere between 8 and 20 billion years old. There is no particular reason why the matter density of a noninflationary universe should have any particular value. The larger range of possible values leads to a larger range of possible ages. This larger range fits better with other estimates: However, since there are so many uncertainties involved, this is not very strong evidence against the inflationary theory. However, estimates of the ages of stars do present problems for the inflationary theory, and sooner or later something is going to have to give.

THE FATE OF THE UNIVERSE

Although there has been something resembling a revolution in the field of cosmology in recent years, ideas about what will ultimately happen to the universe have not changed very much. By the end of the 1970s, scientists had concluded that the universe probably had much less than the critical matter density and that it would therefore expand forever. After the inflationary universe theory became widely accepted, scientists concluded that it was so close to the critical density that it had to be just barely open or barely closed. In either case, it would continue to expand for so long a time that the prediction of when the universe might end didn't seem important.

One thing is clear: Ours seems to be a universe in which matter will become progressively more dispersed. It is destined to thin out until there just isn't very much left to it. It may be that matter itself will eventually decay. The GUTs, for example, predict that the proton—which had previously been thought to be the most stable of all particles—will eventually undergo decay into other particles. If so, then all matter may eventually be transformed into energy at some point countless billions of years in the future.

Of course, if this does happen, there will be no one here to see it. By this time, all of the stars will either have burned out or will have collapsed into black holes, and it is not likely that the universe will be capable of supporting life of any kind. No one really knows what kinds of life might be possible after so many billion years of evolution. Biological organisms may eventually die out and be replaced by highly evolved computers that have artificial intelligence. However, even machines could not continue to exist forever, because the existence of life seems to depend on flows of energy. Life on earth is supported by the energy coming to us from the sun, for example. And the universe will eventually become something that is cold and dark, with practically no sources of energy left.

This fact doesn't necessarily imply, however, that life will never again exist. If a universe could be created once, after all, and be molded into a habitable environment by an inflationary expansion, then there is no reason why this process could not happen again and again. In other words, if there is one universe, there could conceivably be many universes.

I will have more to say about that particular subject in a subsequent chapter. At the moment, I only mention it to provide some words of comfort to those who may be disturbed by the fact that life in our universe will definitely eventually die out. I don't know why we should care whether or not living organisms will still exist billions of years from now, but for some odd reason most of us apparently do. We are disturbed by the fact that the earth is destined to be vaporized when our sun becomes a red giant in about 5 billion years, and we find it depressing to contemplate that the universe will eventually become lifeless.

ARE WE THE RESULT OF MICROSCOPIC QUANTUM FLUCTUATIONS?

Before I go on, there is one other sidelight of the inflationary universe theory that needs to be mentioned. This has to do with the "bumps" or "ripples" in the microwave background that were observed by George Smoot and his colleagues.

If the inflationary universe theory is correct, then these bumps most likely began as events which took place on the subatomic level, and which were then enlarged to visible size by the inflationary expansion. In other words, the density fluctuations that grew into galaxies might have had their beginning in the quantum world.

When I speak of the "quantum world," I am referring to quantum mechanics, the theory used to describe the behavior of atoms, atomic nuclei, or elementary particles. According to quantum mechanics, the subatomic world is a sea of seething activity. As will be discussed later, random events, called quantum fluctuations, continually take place. Under ordinary circumstances these fluctuations have little influence on the everyday, macroscopic world. There are so many of them that their effects average out.

For example, particle-antiparticle pairs can be created out of energy and then disappear in mutual annihilations. But if these events are not tracked in a piece of scientific apparatus, we are never aware of them. Similarly, the emission of a ray of light from an atom is an unpredictable, random event. But when we look at a light bulb, we see only a steady glow. There are so many atoms in the bulb's filament that we are never aware of the randomness.

If there was an inflationary expansion, then perhaps we are now seeing directly into the quantum world. It is not known exactly how long the inflationary expansion (assuming there was one) continued, or precisely how large the increase in the volume of the universe was. However, it is perfectly conceivable that a region of space that was originally the size of an atomic nucleus could have grown into a region as large as the earth, or even into something much bigger.

It is likely, therefore, that the fluctuations that were observed by George Smoot and his colleagues began as submicroscopic events. And if these bumps and ripples did indeed become the seeds from which galaxies formed, then it appears that we owe our existence to events that were taking place on the subatomic level 8 or 10 or 15 billion years ago.

5

What Is the Universe Made Of?

Most astronomers and cosmologists believe that at least 90 percent of the mass of the universe resides in dark matter. Though no one is sure what this dark matter is, scientists are convinced that it exists, and that it is present almost everywhere in the universe. The argument goes something like this:

1. It is very probable that an inflationary expansion took place. Though there is little or no empirical evidence to support this contention, scientists do not know of any plausible way that a universe with features like ours could have been created if an inflationary expansion did not take place.

2. The inflationary universe theory implies that the matter density of the universe must be exactly equal to the critical density, or at least so close to the critical density that we will never be able to tell the difference.

3. But calculations based on the abundance of deuterium and other light elements indicate that the density of baryonic matter in our universe can be no more than about 6 percent of the critical density.

4. Therefore the greater part of the matter in the universe must consist of something else. In other words, at least 90 percent of the universe is made up of something other than such familiar particles as protons, neutrons and electrons.

It is important to understand that, even if the inflationary universe theory were proved wrong, scientists would still have to conclude that some dark matter existed. As we saw in Chapter 2, astronomers have observed the gravitational forces exerted by unseen matter. Moreover, at least some of the dark matter is almost certainly baryonic.

SOME DARK MATTER IS BARYONIC

The only kinds of baryonic matter that can be directly observed by astronomers are those varieties that either emit substantial quantities of light or other radiation (stars and gas clouds) or that absorb it (clouds of interstellar dust and gas clouds again). And there are certainly many objects in our universe that do not fall into either category.

For example, suppose that some of the primordial hydrogen and helium gas collected into bodies that had masses less than 7 or 8 percent of the mass of the sun. Gravity would cause such objects to contract, and the gas of which they were formed would heat up as it was compressed. But, because of their low mass, temperatures would never become high enough for nuclear reactions to begin, and they would never ignite as stars.

Such "failed stars" are called brown dwarfs. We know that at least some brown dwarfs exist, because we can observe one in our own solar system: The planet Jupiter. Though Jupiter has only about one-tenth of 1 percent of the mass of the sun, it can be considered a brown dwarf because gravity has heated up the gas of which it is composed to some degree, causing it to radiate more energy than it absorbs. We should probably consider ourselves lucky that Jupiter did not turn out to be an order of magnitude or two larger. In that case, ours might be a binary star system, and conditions on Earth would probably be inimical to the evolution of life.

Even if many brown dwarfs existed, however, astronomers would not be able to detect their presence because the radiation that these objects emitted would be too insubstantial. Few of the light rays that travel through the universe would fall on them; they would be surrounded by too much empty space. It is possible that they are there but that we can't see them.

Additional baryonic dark matter might exist in the form of intergalactic gas. Such gas can be detected if it exists in substantial quantities. Cool gas emits radio waves, whereas hot gas emits x-rays. Gas clouds can absorb some of the radiation that passes through them as well. Nevertheless, there might exist some gas clouds which are hot enough that they do not absorb much light but not so hot as to emit x-rays. Most likely, such gas does not exist in large quantities. However, it could constitute part of the baryonic dark matter.

Other forms that baryonic dark matter might take include "failed galaxies" and black holes. There might exist dark clouds of gas with masses similar to those of galaxies in which stars never formed. Alternatively, unknown quantities of matter might reside in dark holes. It is possible to make rough estimates of the number of stars that have collapsed into black holes. However, if black holes of this magnitude also formed early in the history of the universe and not as the consequence of stars dying, there would be no way of guessing how many of them there might be. There is no reason why star-sized black holes could not have been created when the universe was much younger and denser than it is today. If so, there might be numerous black holes that are not the relics of dead stars. And if they exist, there is really no way of estimating how many of them there might be.

It appears that we must conclude that there are many forms in which baryonic dark matter could exist. And, as we shall soon see, there are many possible forms of nonbaryonic dark matter too.

But before our speculations go too far, it is a good idea to examine some of the facts, starting with matter we can see.

VISIBLE MATTER

It is possible to accurately estimate the amount of visible matter that exists in the universe. One way is by measuring the light emitted by observable objects. Astronomers can measure the light emitted by a star or a galaxy, and they can use these measurements to estimate its mass. To be sure, galaxies are made up of many different kinds of stars, some are dim and some very bright. However, it is possible to estimate an average brightness and to relate the quantity of light emitted by a galaxy to its mass.

If one assumes that the Hubble constant is about 50, then the density of visible matter in the universe is about seven-tenths of 1 percent of the critical density. This calculation is not a particularly complicated one. Astronomers know how much visible matter there is; all they need to do is add up the mass of everything they can see. And, as we saw in the previous chapter, the critical density and the Hubble constant are related. If scientists knew one, they could easily calculate the other. If the Hubble constant were to turn out to be approximately 85, then this figure would be even less: About two-tenths of 1 percent.

The Hubble constant and the critical density are related. It should be easy to see why. The Hubble constant, after all, is a measure of how fast the universe is expanding. And the faster the expansion, the greater the quantity of matter that must be present to cause the expansion to eventually halt.

When I previously stated that the quantity of baryonic matter in the universe can equal no more than about 6 percent of the critical density, I was assuming that the Hubble constant was about 50. If the constant turned out to be larger, the amount of mass needed to close the universe would be larger, and the quantity of baryonic matter would be a smaller percentage. For example, a Hubble constant of 85 would drop the above figure to 2 percent.

By now, the reader is probably already muttering that all this talk of different Hubble constants and different percentages of visible and baryonic matter makes matter much too complicated. I agree. In the ensuing discussion, I will assume that the Hubble constant is about 50. This is a "conservative" estimate in that it makes some of the constraints less stringent. A larger figure for the Hubble constant would allow even less of the dark matter that is observed to be baryonic.

WEIGHING A GALAXY

Sometimes the mathematical equations that scientists work with become enormously complicated. In many cases, it is possible to obtain solutions only with the aid of high-speed computers, and even then, approximations must often be made. However, sometimes matters are mathematically much simpler than one would expect. One such case involves the rotation of galaxies.

Our sun revolves around the galactic center much as the earth revolves around the sun. Offhand, one might think that the gravitational interactions between our sun and the remainder of the galaxy would be impossible to compute, since the Milky Way galaxy contains something of the order of 100 billion stars. However, to a high degree of accuracy, the motion of our sun around the galactic center depends only on the amount of mass inside the sun's orbit. The gravitational attractions of stars that lie farther away from the galactic center than our sun tend to cancel one another out.

This fact has enormous importance. It implies that one can actually measure the amount of matter in that part of the galaxy that lies within the sun's orbit, and this is possible because we can measure the speed of the sun's orbit. The situation is analogous to that which exists in our solar system: It is possible to determine the mass of the sun simply by observing the orbital velocity of the earth, or of any other planet. If the sun were heavier than it is, it would exert a greater gravitational force on the planets and whip them around it at faster speeds. If the sun weighed less, the planets would move more slowly. A similar effect can be observed by anyone who ties a weight to a piece of string and spins it around. The greater the force that is exerted, the faster the weight will go.

Our sun lies at a distance of about 30,000 light years from the center of the Milky Way galaxy. (This center is not a single body, of course, but a core of billions of stars; nevertheless, the difference is not significant.) Therefore, determining its orbital velocity allows one to compute how much mass there is in our galaxy at distances of 30,000 light years or less. But why stop there? If this method can be used to weigh part of the galaxy, why not use it to weigh the whole thing? After all, this could be done simply by observing the motion of stars on the edge of the galactic disk. Nor is it necessary to stop there. Why not use this method to weigh other galaxies as well?

This can be done without any great difficulty. If one observes a distant galaxy edge-on, one sees that the galactic rotation causes the stars on one rim to move away from the earth and the stars on the other rim to move toward it (this would be analogous to observing the rotation of an old LP record at eye level). In such a case, light from the stars on one side of the galaxy—the side that is moving away—exhibits an additional redshift, and light from the stars on the other side is blueshifted. Velocities can then easily be determined. And if the galaxy is not seen edge-on, no significant complications are introduced. It is a simple problem in trigonometry to make the necessary corrections.

One advantage of this method is that it is not necessary to confine oneself to examining the light emitted by stars that are actually part of a galaxy. One can also look at the motion of glowing clouds of gas that lie outside a galaxy and orbit around it. Thus it is possible to measure not only the mass inside a galaxy, but also any invisible mass that might surround it.

DARK HALOS

When such observations are carried out, a very surprising discovery is made: Galaxies seem to be surrounded by halos of invisible matter. The farther out astronomers go when they perform such measurements, the more mass they discover. In some cases, it is possible to measure the motion of gas clouds that are four or five times as far from the galactic center as stars on the galactic rim. It has been found that these galactic halos extend at least this far, and probably further.

Studies demonstrate that the halos of dark matter seem to have at least ten times as much mass as the visible stars in a galaxy. Since visible matter and galactic halos add up to about 7 percent of the critical matter density, it is just barely conceivable that the dark halos could be made of baryonic matter. As we found earlier, baryonic matter could exist in sufficient quantities to make up 6 percent of the critical density. And figures of 6 percent and 7 percent aren't very different. The small discrepancy might well disappear.

It would be conceivable that this could happen were it not that other evidence points to the existence of even more dark matter. For example, when astronomers study the motions of galaxies in galactic clusters, they find that even larger quantities of mass must be present to hold the clusters together. Galactic clusters, after all, are bound together by gravity. If a certain amount of mass were not present, there would not be enough gravitational force to keep the galaxies from flying apart.

When the appropriate calculations are carried out, it is found that the amount of dark matter in the universe is at least 20 percent of the critical density. The words *at least* are important here, since this method measures only the quantity of dark matter within a cluster—it gives no information about the amount that might be present in the spaces between clusters. And of course it is perfectly conceivable that the entire universe could be filled with a dark matter "sea." In such a case, galaxies and clusters of galaxies would be "islands" in which the mass of dark matter was higher than the average. It would not be correct to assume that most of the mass of the universe was contained in them.

There does seem to be some evidence that such a sea of dark matter exists. During the late 1980s, observations of numerous galaxies out to distances of around 300 million light years were made with instruments that had been placed in the Infrared Astronomical Satellite (IRAS). The resulting data have been analyzed by at least seven different groups of scientists, who agree that measurements of the motions of these galaxies imply that the matter density of the universe is approximately equal to the critical value. Naturally, one cannot state that the two figures are exactly equal. Not only does the exact value of the Hubble constant remain uncertain, but there are uncertainties in the IRAS measurements as well. However, these observations indicate that there is a great deal more dark matter than the amount which is observed inside galactic clusters.

GRAVITATIONAL LENSES

It is conceivable that the dark matter that exists in the vast spaces between galactic clusters may eventually be mapped. Together with colleagues at Bell Laboratories, Princeton University, the University of Cambridge, the Institute for Advanced Study, the National Optical Astronomy Observatories, and the Toulouse Observatory, AT&T Bell Laboratories astronomer Anthony Tyson has developed methods for observing dark matter by means of the gravitational lens effect.

As we saw in Chapter 2, astronomers sometimes observe multiple images of distant quasars when the light from these quasars bends around opposite sides of an intervening galaxy. However, this is not the only kind of gravitational lens effect that is seen. In 1981, astronomer Arthur Hoag of Lowell Observatory noticed an odd blue arc near a galactic cluster on a photographic plate. Five years later, other astronomers reported seeing similar arcs of light, and theoretical scientists responded by performing calculations to determine if these arcs could be indications that dark matter was present in the clusters.

This indeed turned out to be the case. The light from background galaxies was being bent by dark matter in such a way that the arcs were created. Furthermore, when calculations were done to determine how much dark matter was present, it was found that the total amount of mass in the clusters that caused the gravitational lensing was about ten times greater than the visible mass. Since this result was in agreement with estimates of cluster mass that had been based on observations of galactic motions, it seemed reasonably certain that dark matter must be responsible.

Tyson, who has done a great deal of work on dark matter gravitational lenses, now plans to extend the search by looking for gravitational lens effects in places where no galactic cluster lies along the line of sight. If he is successful, he will have discovered clumps of dark matter in places where no visible objects are found.

Meanwhile, other scientists, such as California Institute of Technology astronomer Roger Blanford, have suggested that it might be possible to go even further. According to Blanford, it might be possible to eventually produce large-scale maps showing the distribution of dark matter over distances of hundreds of millions of light years.

BUT WHAT IS DARK MATTER?

One thing is certain. At least 90 percent, and perhaps as much as 99 percent, of dark matter is not baryonic. As we have determined before, both theory and observation indicate that the mass density of the universe is approximately equal to the critical value. The density of baryons is no more than about 6 percent of this figure, and it could be a lot less. If the Hubble constant turned out to have a value of 100, for example, the 6 percent limit would have to be replaced by one that was close to 1 percent.

Thus it is very easy to conclude that most of the mass of the universe resides in something that is not baryonic, but it is extremely difficult to say what this something might be. There have been numerous different theories about what this 90 or 99 percent of the universe might be made of, and all of them have encountered difficulties. The situation has reached the point where some scientists have begun to suggest that dark matter might not be matter at all.

HOT DARK MATTER

The first candidate for the explanation of what made up dark matter was the neutrino. Around 1980, some experiments were performed which indicated that the neutrino—which had previously been thought to have zero mass—might have a small, finite mass after all. The mass that the experimenters claimed to have measured was not very large; it only amounted to about one thirty-millionth of the mass of a proton. However, there are known to be a vast quantity of neutrinos in the universe, so many that even a tiny mass like that could cause them to outweigh the visible matter in stars and galaxies.

Today, cosmologists classify neutrinos as a form of "hot dark matter." There are theoretical reasons for believing that other particles with near-zero mass might exist. If they do, then their behavior—at least as far as cosmology is concerned—would be similar to that of the neutrino. It is quite reasonable to lump all these particles together in the same category. If worst comes to worst and the various hypothetical particles turn out not to exist after all, then the neutrino will still be an example of hot dark matter.

Here the word *hot* is not a reference to temperature as we ordinarily understand the term, but rather to the fact that neutrinos and other light particles would have emerged from the big bang at velocities approaching the speed of light. *Hot,* in other words, means "fast-moving."

However, these high velocities have created problems for the hot dark matter theory. Very rapidly moving particles could not have formed the gravitational seeds for mass concentrations in the universe. On the contrary, they would have broken up any clumps of matter that happened to form.

This does not necessarily imply that galaxies and clusters of galaxies could not have been created in a universe that was full of hot dark matter. After all, the particles would eventually slow down. However, it does imply that galaxy formation would have taken a very long time. Calculations indicate, in particular, that mass concentrations the size of clusters of galaxies would have been created first, and that they would have broken up into smaller masses later.

And this amount of time conflicts with observations indicating that galaxies were formed fairly early in the history of the universe. During the 1980s and early 1990s, astronomers discovered galaxies and quasars (recall that quasars are believed to be the luminous cores of young galaxies) at progressively greater distances from the earth. Since the quasars were billions of light years away, the astronomers who observed them were looking billions of years into the past, into an era when the universe was young indeed. It now seems apparent that some galaxies already existed when the universe was only about 2 billion years old. There is simply no way that the hot dark matter theory can account for galaxy formation that was so rapid.

To make matters worse, the neutrino—the only hot dark matter particle whose existence has been confirmed—could still turn out to have no mass. The experiments which seemed to indicate that it might have some mass were never confirmed, and recent experimental work has allowed scientists to place strict limits on the amount of mass that this particle might possess. As I write this, the most recent results seem to indicate that a neutrino weighs no more than about one hundred-millionth as much as a proton. The maximum mass that it might possess is therefore much less than that which would give a mass density equal to the critical value. Thus it appears that, even if the neutrino did turn out to have mass, there would have to be some other kind of nonbaryonic dark matter present also.

COLD DARK MATTER

Cold dark matter is the collective name given to particles that would have emerged from the big bang at relatively low velocities. Such particles would have been able to clump together in galaxy-sized masses much more quickly than hot dark matter. And if they did, they would have exerted gravitational forces that would draw visible matter together as well. This seemed for many years a plausible scenario for galaxy formation, and the cold dark matter theory was until recently quite popular among scientists.

Though no cold dark matter particles have been discovered, there are a number of different theories of what they might be. Cold dark matter could consist of neutrinolike particles that were even heavier than protons, for example, or it might be made up of hypothetical particles that have been given names like *photino, wino, zino,* and *selectron.* These are the weakly interacting massive particles, or WIMPs, referred to previously. Alternatively, cold dark matter could be composed of yet another hypothetical particle, called the axion, which would travel very slowly, even though it would presumably be very, very light.

I won't make any attempt to catalogue these particles or to discuss them in detail. After all, none of them has yet made an appearance in any experiment conducted in a particle accelerator. And as theories change, so do hypotheses about the kinds of particles that might eventually be found.

However, it is safe to assume that, as more powerful accelerators are built and new experiments are performed, one or more of these particles will make an appearance. Scientists have been predicting the existence of new particles for decades, and experimenters have succeeded in finding them on numerous different occasions. The neutrino itself provides a good example. Though its existence was postulated by the Austrian physicist Wolfgang Pauli in 1930, it wasn't observed experimentally until 1956.

If cold dark matter particles really exist, they interact with ordinary matter only on rare occasions. It is perfectly conceivable that millions, or possibly even hundreds of millions, of WIMPs pass through our bodies every second. If they have no effect on the baryonic matter of which our bodies are composed, we will never be aware of their presence. They are as harmless as neutrinos, even though they are presumably much heavier.

On the other hand, the gravitational effects of cold dark matter particles could be significant. In particular, calculations indicate that cold dark matter would collect into clumps the size of galaxies first, and that masses the size of galactic clusters would form only later. This is just the kind of theoretical behavior that helps to answer some of the perplexing questions cosmologists have, since it seems to correspond to what actually happened in our universe.

Because of this, the cold dark matter theory became quite popular among cosmologists. In fact, until around 1990 it was the most widely accepted explanation for galaxy formation. However, it eventually ran into trouble too. The basic problem was similar to that which did away with the hot dark matter theory. Astronomers began to discover galaxies and quasars at greater and greater distances. And, the farther away such an object was found to be, the closer it had to be to the big bang in time. Eventually the question had to be asked whether galaxies and quasars had formed so soon after the big bang that the cold dark matter theory could not account for them.

I will have more to say about this matter in the next chapter, when I discuss galaxy formation. For now, I will simply observe that it is not enough to guess what the dark matter in our universe might be. It is also necessary to show that this dark matter can produce the kind of universe that we live in. Since galaxies and clusters of galaxies are the most prominent feature of our universe, a good dark matter theory must be able to explain how they came about. And this was something that both the hot dark matter and cold dark matter theories have failed to do.

SHADOW MATTER

Apparently, dark matter does not consist of neutrinos, or other light particles, or axions, or WIMPs. There must be something else present in it as well. If WIMPs were shown to exist, for example, it might turn out that they constituted part of the mass of dark matter. All that has really been demonstrated is that cold dark matter could not produce the features of our universe by itself.

One might think that once the cold dark matter and hot dark matter theories were eliminated, there would be few possibilities left. However, numerous different suggestions have been made, some of them quite bizarre. But, of course, as I have noted before, calling an idea "bizarre" does not imply that it is false. Sometimes the universe turns out to be an even stranger thing than anyone has imagined.

One of the odder possibilities is that dark matter, or at least a substantial portion of it, might be composed of a material called shadow matter. Some of the superstring theories, it turns out, predict the existence of a substance that would interact with ordinary, or baryonic, matter only through the force of gravity. If such matter did exist, it could neither be seen nor felt; only its gravitational effects could be observed. Shadow matter would not be seen because light is a form of electromagnetic radiation, and matter that was not affected by the electromagnetic force could neither emit nor reflect light. And it could not be felt because it is electromagnetic forces that hold atoms and molecules together. If you tried to grasp a chunk of shadow matter, your hand would pass right through it.

It has been said that one could stand in the middle of a shadow matter mountain or at the bottom of a shadow matter ocean and not know it. I suspect, however, that it is probably quite misleading to describe shadow matter in this way. Though some science fiction author will undoubtedly soon be writing a novel about a shadow matter world that exists parallel with our own, such a possibility is quite unlikely. Though shadow matter particles would interact with one another in some manner, it isn't probable that the laws of nature in the shadow matter world would resemble those of our universe.

If shadow matter does exist, it might consist of nothing more than clouds of particles similar to those that might be created from hot or cold dark matter. One thing that we can be sure of is that there are no shadow matter stars or planets in the vicinity of the earth. If there were, their gravitational effects would certainly be apparent.

THE COSMOLOGICAL CONSTANT RETURNS

I will discuss a number of other dark matter possibilities in the chapter on galaxy formation, since these issues are related. Galaxies were created by the force of gravity, and if the greater part of the matter in the universe is dark, it must be responsible for most of the gravitational forces too.

However, before I conclude this chapter, I would like to discuss the possibility that dark matter might not be matter at all.

Some cosmologists have pointed out that, if Einstein's cosmological constant is not exactly equal to zero, then it could be responsible for forces that would mimic the presence of dark matter. For example, a universe that had a matter density of 20 percent of the critical density could be exactly flat if it had a cosmological constant of 0.8. The effects of a constant of this magnitude would be equivalent to those produced by a comparable quantity of mass.

At first, it might not be so obvious why this should be so. However, once that one remembers that mass and energy are equivalent, matters become relatively clear. If the cosmological constant is not zero, then there are forces acting throughout the universe. If there are forces, then there is energy. A nonzero cosmological constant, in other words, would impart an energy density to empty space. But, according to Einstein's equation $E = mc^2$, matter and energy are equivalent. Consequently, this energy would produce gravitational forces similar to those exerted by matter. In other words, there is a way in which a cosmological constant acts like "extra mass."

It is perfectly conceivable that the total "mass" of the universe could be made up of 20 percent matter and 80 percent cosmological constant, or 10 percent and 90 percent, or some other mixture.

Cosmologists have generally tended to believe that although there was a cosmological constant during the period of inflationary expansion, the constant is exactly equal to zero today. There is no empirical evidence that nonzero constant exists, and scientists don't like to construct theories based on the existence of quantities that cannot be detected. To add a cosmological constant to one's equations simply because the dark matter problem seemed intractable would be a suspect procedure, to say the least.

On the other hand, a cosmological constant presumably did exist at one time, early in the history of the universe. If it hadn't there would have been no inflationary expansion. In the opinion of some scientists, therefore, the matter is at least worth discussing. After all, there is nothing wrong with trying to see what a universe with a nonzero cosmological constant might be like and then checking to see if such a hypothetical universe would bear any resemblance to our own.

When this is done, it turns out that more than one problem is solved. Not only does the dark matter problem disappear, but it is also possible to explain the existence of large structures in the universe.

For example, in 1989 astronomers discovered a huge concentration of galaxies, which they named the "Great Wall," that was approximately 500 million light years long, 200 million light years wide, and 15 million light years thick. And the Great Wall, it appeared, was only one of a large number of massive concentrations of galaxies in the universe.

Gravity alone did not seem sufficient to account for such structures, or for the tendency of galaxies to cluster into sheets and filaments that stretched over distances of tens or hundreds of millions of light years. Gravity would presumably be more effective at creating such "small" structures as galaxies than it would be at making very large ones.

To make matters worse, statistical studies gave results that were very difficult to explain if one assumed that only gravitational forces were at work. When the structure of the universe was examined on very large scales, it was found that the chance of finding a galaxy cluster near another cluster was larger than that of finding a galaxy near another galaxy. This was exactly opposite to the result that would be expected if gravity had created galaxies first.

If there were a nonzero cosmological constant, then the problem would immediately become more tractable. The existence of a constant would create additional forces in the universe and these forces could impart an extra "springiness" to spare, and this springiness could help gravity push matter together into large structures.

Finally, the existence of a cosmological constant would clear up the problems associated with observations of stars that seemed to be older than the universe itself. Since the constant represents a force that would give an outward push to the expansion of the universe, a universe with a nonzero constant would be older than one in which no such outward force existed. Such a universe would have started out more slowly: It would have had to in order to attain the expansion rate that is observed today. If two automobiles are both traveling at a speed of 55 miles per hour and one is accelerating, we know that his one must have been traveling more slowly a while ago. This means that it would take longer for such a universe to reach its present size. If there were a nonzero cosmological constant, the universe could very easily be old enough to contain 16-billion-year-old stars.

It would appear that the cosmological constant theory has everything going for it. Why, then, do scientists continue to resist it? Well, it seems to be another case of a beautiful theory being destroyed by ugly facts. The empirical evidence just does not seem to support the cosmological constant idea. Not only are there no observational data to support it, but there also seems to be evidence that can be used against it. Since a universe with a cosmological constant would have been expanding for a longer time than a universe that lacked such a feature, it would have grown to a bigger size. A bigger universe would contain more quasars, galaxies, and other objects along any line of sight. As a result, more gravitational lens effects would be seen. A universe with a cosmological constant of 1.0, for example, should contain ten to one hundred times as many gravitational lenses as one in which the constant was zero.

According to astronomer John Bahcall of the Princeton Institute for Advanced Study, these extra gravitational lens effects are not observed. Making observations with the Hubble space telescope, Bahcall examined hundreds of quasars and found only one gravitational lens effect. This was not nearly enough for a universe in which a cosmological constant was operative.

The cosmological constant theory seems to lack theoretical underpinnings as well. Scientists know of no way to calculate how large a nonzero cosmological constant should be. If the constant turned out not to be zero, they would be faced with the problem of explaining why it existed, and also why it should have one particular value and not another. But no one has any idea where to begin.

It is true that the inflationary universe theory doesn't have much empirical support either. However, the situation there is entirely different. Scientists have good theoretical reasons for believing that an inflationary expansion took place, and they can perform calculations which tell them how and why it should have. These elements are missing in the cosmological constant theory.

But of course the final verdict is not in, either about the cosmological constant or about the nature of dark matter. And there is no reason why dark matter could not have several different components. There are good reasons for thinking that at least some of it is baryonic. It could be partly baryonic, partly cold dark matter, and partly cosmological constant. Hot dark matter could be part of the recipe as well. Certainly matters would be simpler if the effects of one or another of the ingredients dominated all the others. But physical phenomena are not always as simple as one would wish.

6

Where Did the Galaxies Come From?

Dinosaur Footprints

The discovery of "bumps" or temperature fluctuations in the big bang created waves of excitement in the scientific community when it was announced in the spring of 1992. (See Chapter 1.) Before long, scientists were saying that this new discovery would usher in a golden age of cosmology. Now that the fluctuations had been detected, a whole new series of discoveries was bound to follow. New fields of research would open up, and ideas about the structure and evolution of the universe were bound to be revolutionized.

Some scientists expressed themselves in especially colorful ways. "It's our version of a dinosaur footprint," said John Mather of the Goddard Space Flight Center, one of the members of the COBE team. The implication was, of course, that scientists had found a relic of the distant past, one that they had been seeking for well over a decade. They no longer had to theorize about the kinds of density fluctuations that had existed in the early universe.

The analogy of a dinosaur footprint may have been even more apt than Mather intended. The discovery of a fossil footprint, after all, does not allow paleontologists to reconstruct the appearance of the entire animal. At best, it only gives information about the creature's size and gait. Similarly, although the discovery of ripples in the big bang was a momentous event, it provided surprisingly little new information about the structure of the universe. In particular, the outstanding problem of cosmology—the question of how galaxies were created—remained as mysterious as ever.

The fluctuations detected by the COBE observations corresponded to structures that were much larger than galaxies, indeed much larger than clusters and superclusters of galaxies. But this glimpse of the early universe did not enable scientists to study the early cosmos in fine detail.

At first scientists thought that the COBE data did provide the basis for ruling out some popular theories of galaxy formation. The proponents of the leading theories immediately set to work, and in most cases they discovered that their hypotheses could be modified in ways that could accommodate the new results. Knowledge about conditions in the early universe is so indefinite and incomplete that these theorists had to do little more than change a parameter here and modify a numerical estimate there. In the end, however, scientists were left with almost as many competing theories of galaxy formation as they had before George Smoot and his colleagues announced their discovery.

At this point, the reader may indignantly ask, "*This* was how the new era in cosmology began?" My answer would be "Of course. Why not?" Isn't this how golden ages of science are ordinarily inaugurated? The great ages in science are not those in which all the answers are known (or in which it is thought that the answers are known). The great ages are the ones in which scientists grapple with deep and baffling problems. They are ages in which the concerted efforts of hundreds or thousands of scientists lead to new discoveries that allow these problems to be solved, bit by bit.

DARK MATTER AGAIN

The problem of galaxy formation and that of dark matter are closely related. Indeed, as I noted in the previous chapter, certain ideas about the nature of dark matter, such as the hot dark matter theory, were eventually discarded because they could not adequately explain the process of galaxy formation.

For a number of years, the leading theory of galaxy formation was the cold dark matter theory. Although this theory turned out to be inadequate in its pure form, it is still worth discussing in some detail. Although cold dark matter alone could not have produced the number of galaxies we observe in the amount of time we estimate they were formed, galaxy formation could have been seeded by clumps of cold dark matter working in cooperation with something else. The idea that cold dark matter is the only type of nonbaryonic dark matter has been discredited. However, it is still possible that cold dark matter is a part of the mix.

In its pure form, the cold dark matter theory was based on the hypothesis that gravity caused cold dark matter particles to clump together, and that galaxies were created when the primordial hydrogen and helium gas collected around these clumps.

This theory was in trouble even before the COBE results were announced. The reason was that, even though galaxies could be created around cold dark matter more quickly than they could around hot dark matter, they couldn't be created quickly enough. As I have previously explained, during the late 1980s and early 1990s, astronomers observed galaxies and quasars at greater and greater distances. Finally, they found some at such early eras that the pure cold dark matter theory could no longer be modified to account for them.

QUASARS

Much of the evidence against the cold dark matter theory is based on observations of quasars, which are the oldest objects—aside from the cosmic microwave background—that can be observed in our universe. Since astronomers gain important information about the young universe by observing quasars, we should discuss currently accepted ideas about the nature of quasars before going on.

The word *quasar* is an abbreviation for "quasi-stellar radio source." When quasars were first observed during the early 1960s, they were detected because they were such strong emitters of radio waves. Scientists soon discovered that these radio emitters could generally be associated with visible objects that looked less like galaxies than like stars. This is why they were called "quasi-stellar."

When the redshifts of the quasars were measured, however, the quasars were found to be billions of light years from the earth. This implied that they did not resemble stars at all. No star could be this bright. The very fact that quasars could be observed at such great distances indicated that they were even brighter than the most luminous galaxy. The quantities of light that they were emitting were enormous. Some of them had the luminosity of trillions of stars.

And yet quasars appeared to be relatively small. Astronomers found that their diameters were not much greater than that of our solar system. Naturally, the size of quasars could not be measured directly. Such fine detail cannot be observed at such great distances because anything that is billions of light years away will appear as a point of light even in the most powerful telescope. However, the light output of quasars often changes abruptly, and this allows certain deductions to be made about their dimensions.

In some quasars, the quantity of light emitted changes noticeably over a period of only months, or even days. It is therefore possible to conclude that they are much smaller than galaxies. The reasoning goes something like this: Suppose that there existed some kind of disturbance that could alter the light output of an object the size of a galaxy. Since, at best, such a disturbance could propagate no faster than the speed of light, it is obvious that tens of thousands of years would have to pass before the object's luminosity could change appreciably. It would take that long for the disturbance to travel any appreciable distance across the object.

Studies of variations in quasars' luminosities indicated that quasars were probably no more than about a light day in diameter. This is about twice the size of our solar system; it would take a ray of light about half a day to travel from the orbit of Pluto, go past the sun, and then travel out to Pluto's orbit again.

At first, some astronomers would not believe that so small an object could produce so much energy. They suggested that the apparent great distance of quasars might be an illusion, that these objects could be much nearer—and therefore much dimmer—than they seemed to be. However, as the astronomical evidence accumulated, the theory became generally accepted that quasars were indeed very distant objects that had existed early in the history of the universe.

A number of theories have been invented to explain the great luminosity of quasars. According to the one considered most plausible, a quasar is a giant black hole with a mass 100 million (or more) times greater than that of our sun, which is situated in the core of a newly formed galaxy. There are some other theories which seek to explain the luminosity of quasars without resorting to the assumption that a black hole is present. But, since they generally begin with the assumption that the galaxy contains a compact central mass of some kind, they are not as different from the black hole theory as one might think.

BLACK HOLES ARE BRIGHT

A black hole can emit no light. Nor can it reflect any, since it absorbs any matter or radiation that comes too near to it. As a result, laypeople sometimes conclude that black holes must always be dark objects. In reality, they are nothing of the sort.

If there is any gas in the vicinity of the black hole, it will be accelerated by the hole's intense gravitational field. Gas particles will spiral around the black hole and move faster and faster until they finally fall into it and are absorbed. As they spiral in, they will emit radiation of various frequencies, including light and x-rays. Astronomers have discovered several objects they believe are (stellar) black holes because of the nature of their x-ray emissions.

The idea that quasars are powered by massive black hole "engines" in young galaxies is very attractive. One would expect, after all, that a young galaxy would contain a great deal of interstellar gas. As this gas fell into the central black hole, it would be accelerated to such a degree that the light output would be enormous.

If this theory is correct, one would expect to find supermassive black holes in the cores of many, or possibly all, galaxies today. There is indeed some evidence for this. For example, some observations suggest the presence of a massive black hole in the center of our own galaxy. And there is evidence indicating that a body of

10 million solar masses or more exists in the center of the great galaxy in Andromeda.

Finally, it is not difficult to imagine how such supermassive black holes would have formed in the first place. The stars that make up a galactic core are very closely packed. The more massive of the stars in the center of a young galaxy would have exhausted their nuclear fuel and have evolved into black holes fairly quickly. It is reasonable to assume that, once these black holes had formed, they would have absorbed much of the interstellar gas in their vicinity, and possibly also stars whose orbits brought them near to the black holes' surface. As the black holes absorbed quantities of matter, they would have grown larger and have begun to attract nearby matter even more strongly. Sooner or later, some of these black holes would merge with one another, eventually creating a supermassive object. Once it was created, it would continue to swallow nearby stars and gas and grow larger yet.

Obviously, it is not possible to look at events taking place in the cores of young galaxies during the first few billion years after the big bang in order to see if this is what happened. However, this is an extremely plausible scenario. In fact, it is difficult to imagine what could have prevented such a series of events from taking place.

QUASARS AT HIGH REDSHIFT

At this point, it is necessary to go into a technical matter. As we have seen, astronomers do not know precisely how old the universe is, or how far away many of the objects they observe really are. As a result, they habitually speak of an object's distance (and age) in terms of redshift. Instead of saying that a quasar is so many billions of light years away or that they are looking that many billions of years into the past when they observe it, they might say something like "This quasar lies at redshift 3.8."

Such statements are more meaningful than one might think. They state explicitly how much the universe has evolved since a quasar (or some other object) emitted the light that is observed today by terrestrial telescopes. When scientists say that an object lies at a redshift of 1, they mean that the light waves by which it is observed have been redshifted or "stretched" by a factor of

100 percent. This measurement can also be interpreted to mean that the dimensions of the universe are 100 percent greater now than they were then (galaxies are twice as far apart). Similarly, if an object lies at a redshift of 2, this means that it is much farther away; its light has been "stretched" or redshifted by a factor of 200 percent.

Calculating the age of an object with a given redshift is a little more complicated. The figure that one obtains depends not only on the redshift but also on whether the universe is really at critical density. It also depends on whether there is a cosmological constant. However, astronomers can use redshift measures to make some determinations about the age of the object observed. They know that when they observe an object of redshift 1, they are seeing it as it was when the universe was *less* than one-half its present age. The reason for this is that the universe was expanding faster in the past than it is now. Therefore it must have taken less than half the total elapsed time to reach the halfway point in its expansion. The situation is similar to that of a car that is being braked. Since it is slowing down, it will take a longer time to travel the second half of the distance it must traverse before it comes to a halt.

Most quasars are seen at redshifts ranging from 1 to around 3. But there are some that are even farther away. In 1982, for example, a quasar was found at redshift 3.78. It was so far away that its speed of recession was more than 90 percent of the velocity of light, which means that the light by which astronomers observed it had been emitted when the universe was only about 10 percent of its present age.

The cold dark matter theory of galaxy formation could accommodate a date as early as this for the first galaxies, but just barely. According to the computer models used to test the theory, it was apparent, however, that the cold dark matter scenario for galaxy formation would be in trouble if quasars were found at redshifts greater than 4.

And, unfortunately for the theory, this is exactly what happened. In the one year between August 1986 and September 1987 alone, astronomers discovered, not just one, but seven quasars with redshifts between 4 and 4.4. Since that time, they have found objects that were even farther away. As I write, the record for a quasar redshift stands at 4.9. This particular quasar already existed when the universe was only about 7 percent of its present age. If the universe is indeed about 15 billion years old, then one must conclude that this quasar formed within approximately a billion years after the big bang.

A billion years is certainly a long time in human terms: about 500 times longer than the genus Homo has existed. However, it is not a very long interval when one is speaking of the evolution of the universe or the formation of quasars and galaxies. As of current knowledge, there is simply no way that the cold dark matter theory of galaxy formation can accommodate such a figure without drastic alterations. Thus, in the eyes of many cosmologists, the theory is dead.

As I have already pointed out, the fact that the theory has been discredited doesn't necessarily imply that cold dark matter does not exist, only that it alone cannot be responsible for galaxy formation. As I write this, the idea that galaxies might have been formed around clumps that were a combination of hot and cold dark matter has become a fashionable hypothesis. It remains to be seen, however, whether this theory will be able to explain all the observed phenomena, or whether it too will eventually be discarded.

COSMIC STRINGS

According to another theory, cosmic strings might have played a role in galaxy formation. Cosmic strings have nothing to do with the superstring theories I have mentioned from time to time. In fact, cosmic strings, as opposed to superstrings, would not be microscopic objects. On the contrary, they would be massive structures that might easily have seeded galaxy formation, if they existed.

According to the GUTs and certain other theories of elementary particles known as supersymmetry theories, space can exist in a number of different energy states. Indeed, as we have seen, the inflationary universe theory makes use of this idea. The inflationary expansion presumably took place as the universe underwent a transition from one state of energy to another.

Such transitions are called phase transitions, and they are analogous to certain sudden transitions that we can observe in the everyday world, such as the freezing or boiling of water. Water and ice, for example, have the same chemical composition. However, their physical properties are very different because they have undergone a transition from one state to another.

When water freezes, moreover, defects can appear in the crystal structure of the ice that is formed. For example, one often sees cracks in the surface of a frozen lake. A flaw in a diamond is a similar phenomenon.

A cosmic string would be something very similar to these flaws. It would be the discontinuity in the structure of spacetime that was created when the universe underwent a phase transition, presumably at the time of the inflationary expansion. If any cosmic strings exist today, they would have the form of long, filamentlike concentrations of energy. They would bear certain similarities to those hypothetical particles known as magnetic monopoles, which also would have been created during a phase transition. The difference between them is that a monopole would resemble a mathematical point, whereas a cosmic string would bear a greater resemblance to that crack in a sheet of ice.

If cosmic strings exist, then they must be very massive. After all, they would be relics of an era when energies were enormous. It has been calculated that a piece of cosmic string the size of an atom would weigh a billion tons, and that a section long enough to stretch across a football field would weigh as much as the earth. Cosmic strings have never been observed. However, if they are real, they could very well have played a role in galaxy formation. After all, they represent concentrations of mass that could have been the gravitational seeds around which matter condensed.

Cosmic strings would not stay in one place for long. On the contrary, they would move about at nearly the speed of light. However, if there were strings in the early universe, they would have had a tendency to twist about and intersect with one another. If this happened, some of them would have formed closed loops. These loops could have seeded galaxy formation, possibly in cooperation with hot dark matter. If neutrinos, a possible component of hot dark matter, have a nonzero mass, for example, the gravitational attraction of strings would have caused these particles to cluster together at an earlier epoch than they would have if the cosmic strings had not began present.

Even if cosmic strings once existed in great numbers, there might not be many of them left today. Theory predicts that cosmic strings would radiate their energy away and eventually disappear. The smaller strings would evaporate relatively quickly. The larger ones would last longer, but they would eventually disappear also. If this scenario is accurate, it almost seems that cosmic strings were created for the express purpose of making galaxies, and that they exited from the stage as soon as their job was done.

Even if no cosmic strings exist today, it might still be possible to determine that they once did exist. After all, astronomers habitually look billions of years into the past. There is no reason why they could not observe a gravitational lens effect created by a cosmic string that existed at some suitably high redshift, in other words, at the theoretically appropriate age after the big bang. It might even be possible to see cosmic-string-induced irregularities in the cosmic background radiation. As I write this, however, no effects from cosmic strings have been seen, so the idea that they might be responsible for galaxy formation is still highly speculative.

ANY OTHER IDEAS?

As a matter of fact, there have been numerous other suggestions about possible mechanisms for galaxy formation, some of which might contribute to solving the dark matter problem as well. As we have seen, most cosmic strings would probably have disappeared by now. Consequently, even if they did once exist, they are not likely to be a significant component of dark matter today. However, this is not the case when one considers some other possible scenarios.

If one wants to come up with new ideas of how galaxies might have been created, a good place to start is to ask what kinds of dark matter could conceivably exist that were not discussed in Chapter 5, and that could account for galaxies forming. It turns out that there are several interesting theoretical possibilities.

The universe could contain objects known as quark nuggets, for example. No one really knows whether quark nuggets exist, but there is no good reason why they shouldn't. If they do, they could contribute significantly to the average mass density of the universe and yet remain unobserved.

Quarks are the theoretical components of protons, neutrons, and other heavy particles (on the other hand, electrons, neutrinos, and electronlike particles such as the muon are not made of quarks). Quarks are never observed in a free state. Numerous attempts have been made to detect isolated, unbound quarks, and all have failed. But quarks have been detected inside baryons. For example, in 1968, an experiment was performed at the Stanford Linear Accelerator Center (SLAC) in which it was found that there existed small, pointlike charges inside protons, just as the theory had predicted. Since then, experiments have confirmed this result on numerous occasions.

It is likely that, during the early stages of the big bang—and here I am speaking of a time after the inflationary expansion but before the creation of helium and other light elements—free quarks did exist. The universe was very hot at this time, and quarks and antiquarks would have constantly been created out of pure energy in vast numbers.

As the universe expanded, it cooled rapidly, and when the universe was about a millionth of a second old, the quarks combined into protons, neutrons, and other particles. But there may have been some that did not, and that instead might have condensed into quark nuggets, dense chunks of matter that might have ranged from about a tenth of a centimeter (about a twenty-fifth of an inch) to about one meter (approximately a yard) in diameter.

Calculations indicate that if quark nuggets do exist, they are likely to contribute to the average mass density of the universe to about the same extent that baryons do. Thus they could conceivably play a significant role in galaxy formation, and perhaps account for the theorized amount of nonbaryonic matter as well. This makes them quite an appealing possibility. Unfortunately, there is as yet no evidence that quark nuggets did ever form, and they are at best an exotic possibility.

PRIMORDIAL BLACK HOLES

Another possible explanation of galaxy formation involves primordial black holes. It is conceivable that some or all of the nonbaryonic dark matter could consist of primordial black holes, and if so, they could have played an important role in galaxy formation. In fact, there is no reason why such black holes could not constitute a significant part of the dark matter that is observed in galaxies.

When I speak of primordial black holes, I am not talking of objects that are the remnants of collapsed stars, but rather of black holes that might have been created when the universe was only a fraction of a second old, long before the era of star formation. Under the extreme conditions that existed then, matter and energy could have been compressed enough to produce black holes here and there.

Nor would primordial black holes necessarily have masses similar to those of black holes that are stellar remnants. In fact, any black hole that was created before the universe was a thousandth of a second old would have a mass much less than that of our sun. Theory indicates that there is a relationship between the mass of such a black hole and the time at which it was formed. Some of them could be tiny indeed, at least when compared to the black holes that are created from stars.

But do primordial black holes exist? If they did it would be difficult to observe them directly. For example, a black hole with a mass of 10 billion tons would have a diameter that measured only about a hundred-billionth of a millimeter. Black holes that were created at about the time that quarks combined to form baryons and other particles would be considerably larger. They would have masses about a thousandth of that of our sun and would be about the size of footballs. But it would not be possible with current technology to see them either. No one has thought of any way yet to find football-sized objects scattered through the dark reaches of the universe.

Though primordial black holes cannot be detected by astronomical observations given our current technology, it is conceivable that scientists might eventually obtain evidence of their existence. The British physicist Stephen Hawking has shown that it is theoretically possible for a very small black hole to disintegrate. At the moment of its disruption, such a black hole would produce bursts of high-energy gamma radiation. If such radiation was detected, and if it could be shown that it had indeed come from dying black holes, then one could not only confirm that primordial black holes existed but could also calculate their contribution to the mass density of the universe. However, at the moment, like cosmic strings and quark nuggets, primordial black holes must be considered nothing more than an interesting theoretical possibility.

EXPLOSIONS, LATE PHASE TRANSITIONS, AND TEXTURES

One theory suggests that galaxies are not created by gravitational seeds at all but are the result of gigantic explosions. According to this theory, loops of cosmic string might behave in a manner similar to the superconducting materials that scientists have produced in laboratories. Subatomic particles trapped inside such superconducting strings could conceivably produce enormous electrical currents. For example, electrons could travel around a loop of string much like electrons travel down a length of wire.

Every electrical current gives rise to a magnetic field. This is the principle on which electric motors operate. In electric motors, the magnetic field produced by a coil of wire causes magnets to move. The moving magnets provide the power to run a vacuum cleaner, or a coffee grinder or a subway train.

If cosmic strings once existed, they might very well have behaved as superconductors. This means that once an electrical current was set up in a string, no additional energy would have been required to maintain the current. Meeting no resistance, the electrons and other charged particles trapped in the string would have continued to travel inside the string for an indefinite period of time.

This would have created electric and magnetic fields that were enormously strong. In turn, these fields might have interacted with the hydrogen and helium gas that filled the universe to create rapidly expanding bubbles of hot gas. In other words, the superconducting strings might have produced gigantic explosions. A terrestrial explosion, after all, is nothing but an expanding shock wave, and the bubbles of gas would have constituted shock waves on a cosmic scale.

Galaxies and clusters of galaxies would then presumably have been created when bubbles expanding in different directions came into contact with one another. The waves of compression that were created when bubbles met would have compressed any matter that was caught between them.

Another theory that has been posited is premised on the possibility that the phase transitions that presumably changed the character of spacetime were not confined to the first fraction of a second after the big bang. The late phase transition hypothesis begins with the assumption that such a transition might have taken place millions of years later. If so, the transition would have created defects in spacetime which were not microscopic structures like monopoles and cosmic strings (a cosmic string might be very long, but it wouldn't be very thick), but rather "domain walls" millions of light years thick. These walls would separate different sections, or domains, of the universe in which space had the same characteristics that it exhibits today.

According to this theory, the walls might then have broken into chunks, and these chunks, or "balls of wall," might have seeded galaxy formation. Alternatively, antigravity forces created by a pair of walls could have caused matter between them to be compressed.

Unfortunately, this late phase transition theory seems to be ruled out by the COBE results. The temperature fluctuations found by Smoot and his colleagues seem to indicate that significant density fluctuations already existed when the universe was 300,000 years old. In other words, the clumps of matter that would have seeded were already there before the late phase transitions presumably took place.

But theories can always be modified, and it is not inconceivable that the late phase transition theory, though highly controversial, could be reborn in a new form. When one is dealing with extremely speculative theories that try to wrestle with phenomena so little understood as galaxy formation, one never knows what is going to happen.

There exists yet another theory which connects galaxy formation to phase transitions, though it is highly controversial. This theory postulates the existence of objects known as textures. Like cosmic strings and domain walls, textures would be defects in spacetime created by phase transitions. However, they would have a somewhat more complicated form. It has been shown that it is mathematically possible that they might contain twists analogous to those in a twisted rubber band. According to this theory, if textures are real, they could explain, not only galaxy formation, but also the existence of even larger structures in the universe, such as clusters.

Shortly after the COBE results were announced, the claim was made that the texture theory was now ruled out. However, the proponents of the theory quickly mounted a counterattack, and as I write this, the matter is controversial.

PEEBLES' DIG AT HIS COLLEAGUES

One theory of galaxy formation is unique in that it does not depend on the existence of various kinds of different exotic, undiscovered objects. Proposed by Princeton University astrophysicist P. James E. Peebles, this theory makes use only of materials that scientists *know* were present in the early universe: Baryonic matter and radiation.

During the course of a phone conversation that I had with him, Peebles admitted that his theory was intended partly as a dig at his colleagues. Although it is a serious hypothesis, it manages to poke a little fun at theoretical scientists who try to explain galaxy formation by inventing all sorts of exotic mechanisms.

Nevertheless, the theory is not a joke. Peebles is one of the leading authorities in the field of cosmology, and any theory that he proposes—whatever its motivation—has to be taken quite seriously. Furthermore, the hypothesis is "about consistent" (as Peebles put it) with the COBE results. No drastic modifications have to be made in it to accommodate the new data.

Peebles assumes a low-density universe in which the only matter that exists is baryonic. The rationale for this premise is a simple one: Though scientists have theorized about nonbaryonic matter endlessly, the baryonic variety is the only kind that has ever been seen.

The baryonic matter in Peebles' universe is assumed to have a density that is about 20 percent of the critical density. As we have seen, scientists generally think that about 6 percent is a reasonable maximum figure. However, given all the uncertainties that one encounters in the fields of astronomy and cosmology, a figure of 20 percent is not impossible.

Peebles' theory seems to imply that we live either in an open universe that will expand forever or in a flat universe with a cosmological constant. After all, there is no nonbaryonic dark matter that would account for the 80 percent "deficit" in the mass density. The universe would not contain enough mass to be closed. Peebles is not particularly comfortable with the idea of a cosmological constant. Like many other scientists, he emphasizes that there isn't a shred of empirical evidence to indicate that it has any value other than zero. And of course, if there is no cosmological constant and the theory is correct, then there can have been no inflationary expansion either. As has been pointed out before, an inflationary expansion would create a universe in which the average matter density was almost precisely equal to the critical density.

Peebles is not trying to "prove" that the inflationary expansion did not take place (he does point out that there is no empirical evidence for such an event, however). He is simply presenting a hypothesis that is based on observation, and nothing but observation, and trying to see where it leads. If his theory continues to be consistent with observed data, of course, the possibility of a noninflationary universe might be something to consider. However, at the present time—and I think that Peebles would agree with me—any attempt to discard the inflationary theory would be quite premature.

According to Peebles' theory, the process of galaxy formation was simplicity itself. He postulates that, when the universe was about 100 million years old, the primordial matter that was present broke up into gas clouds. At first, these clouds filled the entire universe and the universe began expanding. But then, as the expansion continued, the clouds began to move away from one another, simultaneously contracting as their temperatures dropped. As their contraction continued, star formation began. By the time the universe was about half a billion years old, a star-filled sky was ablaze with light.

In Peebles' theory, galaxies would have formed in plenty of time to have produced quasars at redshift 4.9. The theory also explains that the vast spaces that separate galaxies in the present-day universe are a result of the galaxies' early formation. And of course there is also time, according to Peebles, for gravity to have caused galaxies to cluster together into the large structures that astronomers observe today.

Of course, you never get something for nothing. Though Peebles' theory is beautifully simple, it has drawbacks. If we are to believe that galaxies were created in the manner that he envisions, we must assume that there were fluctuations in the numbers of baryons that were present in these different locations in the early universe, since galaxy creation requires that some regions are denser than others. These fluctuations would be balanced by variations in temperature. In places where baryons were less abundant, they would be moving more rapidly.

The necessity of making this kind of assumption is a drawback, since it is not obvious why such fluctuations in baryon numbers of the particular size required should have existed. Nevertheless, the theory is quite appealing, for it seems to show that a universe with features like those that astronomers observe could have been created with a minimal number of ingredients. And it encourages the hope that it might yet become possible to explain all the details of galaxy formation without having to resort to assumptions about exotic objects which might not even exist.

7

WHAT IS NOTHINGNESS LIKE?

Contemporary physicists believe that there are deep connections between the theories which describe events that take place in the microworld and those which describe the evolution of the universe. The reason is very simple. As we have seen, contemporary theories of cosmology have to account for phase transitions and other processes involving the nature of space itself. So if we are to be able to discuss the character of "nothingness," that is, empty space, it will be necessary to know something of quantum mechanics, which is a theory of the behavior of subatomic particles.

One of the basic principles of quantum mechanics is the uncertainty principle, which was formulated by the German physicist Werner Heisenberg in 1927. When Heisenberg proposed this idea, he was trying to solve certain outstanding problems in atomic physics. He had no idea that his conception would eventually be used in theories that attempted to describe the nature of the universe itself.

According to Heisenberg, the position and momentum of a subatomic particle could never be simultaneously determined.

The more accurately one knew one quantity, the less accurately one would know the other. For example, if an electron could be precisely located at some point in space, then its momentum, or its velocity (since momentum is mass times velocity), would be completely unknown. Alternatively, if the electron's velocity (or momentum) could be determined exactly, then it would not be possible to obtain any information about its position at any given moment of time. Finally, if one's determination of either position or velocity were a little fuzzy, then one's knowledge of the other quantity would be uncertain to a corresponding degree.

Momentum cannot be measured if position is known because, in such a case, it cannot even be *defined.* If position is known exactly, the very concept of the "momentum" of the electron loses all meaning. This is the real meaning of the "uncertainty" principle.

Similarly, if the momentum (or velocity) of an electron could be determined exactly, then it would not be possible to assign the particle to a certain position in space. The electron would no longer be in this atom, or in that one. Its position would have dissolved into probability waves that spread out over all of space.

Depending upon the precise definition of *uncertainty* that is used, there are several alternative ways of expressing the uncertainty principle mathematically. For our purposes, the following expression should serve:

$$\text{Uncertainty in position} \times \text{Uncertainty in momentum} = \text{Constant.}$$

Since the product of the two numbers is always the same, it follows that when one of the quantities grows smaller, the other must become larger, and vice versa. The situation is somewhat analogous to the payment of overtime salary. If a bus driver gets double time for working on Sunday, he obviously only has to work half as many hours that day to earn a certain number of dollars.

The constant that is the product of the uncertainties is called Planck's constant, after the German physicist Max Planck, who first used it to describe the character of light from black bodies in 1899. Planck's constant is equal to 6.6×10^{-27} erg seconds. If you don't know what an erg second is, don't let it bother you. An erg is a small unit of energy, but that isn't the important thing. What is significant is that this is a very small number: If it were written in decimal form, there would be twenty-six zeros between the decimal point and the first digit 6. If Planck's constant is this small, then the uncertainties in position and momentum (or velocity) must be very small too. As a result, Heisenberg's uncertainty principle is only significant when one is dealing with subatomic particles.

Theoretically, the principle should apply to the macroscopic world too, but in practice it doesn't. If the position of a billiard ball has an uncertainty of a trillionth of a centimeter, this fact is of no practical importance to the player, since such a small quantity cannot even be measured. On the other hand, an uncertainty of a trillionth of a centimeter in the position of a proton could be quite important. The strong force—the force which binds protons and neutrons together—has a range which is an order of magnitude less than that. And the weak force, which also affects protons, has a range which is even less. Moreover, one can even say that the uncertainty of the proton is greater than that of the billiard ball, since, compared with a proton, a billiard ball has a very large mass. Since mass is part of momentum, the corresponding uncertainty in position would be much less than it would be for the much lighter proton.

When one is dealing with objects like a billiard ball, then, it is perfectly reasonable to assume that the effects of Heisenberg's principle don't apply. Nor is it possible to discern any effects the principle might have on objects the size of dust particles, or bacteria, or cells in the human body. Perhaps we're fortunate. Imagine the confusion that might result if the uncertainties were truly large. For example, the game of baseball would be rather difficult to play if umpires could not tell whether pitches were balls or strikes, and shortstops were unable to determine whether ground balls were bounding in an upward or a downward direction. For that matter, imagine the confusion that would result if the uncertainties were really large, and it was impossible to tell whether the brain cells thinking the thoughts I am now writing were in your head or mine.

That idea, however, is not as bizarre as it sounds. To be sure, I can be reasonably certain that my cells remain within my body. However, a similar situation does not exist in the case of a small particle such as an electron. It is often impossible to say whether an electron is in one atom or another. In such a case, the best that one can do is to assign a certain probability to its being found in each.

QUANTUM UNCERTAINTIES

The uncertainty principle is not something that simply places limits on the accuracy to which certain quantities can be measured in the laboratory. If it were, it would do nothing more than provide scientists with certain rules of thumb that they could follow when trying to determine the theoretical precision to which certain quantities could be measured. But Heisenberg's principle has a significance that is much deeper than that. In fact, its implications are so profound that scientists are still discussing them more than sixty years after the principle was proposed.

It is the uncertainty principle which gives quantum mechanics its paradoxical character. Since, as we have seen, under certain circumstances the concept of the "position" of a particle no longer applies, a subatomic particle can sometimes be in two different places at the same time. In fact, experiments have been performed in which a neutron or a photon (a photon is a particle of light; another implication of quantum mechanics is that light can be a wave and a particle at the same time) follows two different paths through a piece of scientific apparatus. The particle does not split into two parts in order to accomplish this; it is literally in two different places at once.

The picture that quantum mechanics give us of physical reality is a strange one. Scientists have been discussing and arguing about the theory for more than half a century, and they generally concede that they are nowhere near understanding all of its implications. But they generally agree that quantum mechanics seems to be a correct theory. When one stops trying to picture precisely what it is that subatomic particles are doing and uses the theory to make numerical predictions that can be tested by experiment, these predictions are almost always confirmed to a high degree of accuracy. Quantum mechanics is a theory that can be used, not only to describe the behavior of subatomic particles, but also the emission of light from atoms, the behavior of transistors and other electronic devices, the phenomenon of radioactive decay, the nuclear reactions that take place within the sun, and many other phenomena.

Moreover, even though quantum mechanics tells us nothing very useful or interesting about the behavior of an object as large as a brain cell or a bacterium, it has profound implications for our understanding of the universe. As we have already seen, work in cosmology and work in particle physics are very much bound up with one another. For example, the idea of an inflationary expansion was suggested by the grand unified theories, which, like most theories in modern physics, are based on quantum mechanics. Conversely, the field of cosmology has become a kind of laboratory for high energy particle physics these days. The energies possessed by subatomic particles during the early stages of the big bang far exceed any that will ever be attained in terrestrial particle accelerators. Therefore, new discoveries in astronomy can sometimes be important to those physicists who study the quantum world.

In some cases, results in cosmology and those in the field of high energy particle physics corroborate each other. For example, it is possible to relate the number of different kinds of neutrinos to the expansion of the universe. If there had existed more than three different varieties of neutrino during the early stages of the big bang, the extra neutrinos would have carried away energy and the expansion rate of the universe would have been altered. One can therefore use findings in cosmology to argue that no more than these three different varieties can exist. Interestingly, at approximately the same time that this result was announced, particle physicists were performing experiments which also indicated that the number of different kinds of neutrinos could be no greater than three.

THERE IS NO SUCH THING AS NOTHING

Position and momentum are not the only quantities that cannot be determined simultaneously. There are actually a number of different quantities that can be paired in this manner among them, energy and time. This connection is a little more difficult to understand than that involving position and momentum. After all, "time" is not, strictly speaking, a property of a particle. On the contrary, a particle exists *in* space and time.

Nevertheless, the application of the uncertainty principle to the energy-time pair yields some important results. For example, if one knows the energy state of an atom exactly, then it is impossible to say how long the atom will remain in that state. The same considerations apply to the energy of atomic nuclei. This is why radioactive decay is such an unpredictable process. One simply cannot say when a given nucleus is likely to emit an alpha particle, for example. The Heisenberg principle prevents us from doing so.

But, as I pointed out a little earlier, the uncertainty principle is not something which stops at placing limits on human knowledge. To be sure, it does do that. But it also goes much further. It tells us that the nature of reality on the subatomic level is something quite different from the nature of reality in the macroscopic world. When one is dealing with the very small, processes can take place which seem bizarre indeed.

For example, the uncertainty principle implies that it should be possible to create matter out of nothing. As you recall from Chapter 4, whenever matter is created, it appears as particle-antiparticle pairs. The creation of such pairs is a process that has been observed many times in the laboratory. When sufficient energy to create them exists, the particle and antiparticle will generally fly apart from one another. The particle will continue to travel through space, and nothing very extraordinary will happen to it. The antiparticle, on the other hand, will undergo annihilation as soon as it encounters a suitable particle of matter. An antiproton will only continue to exist until it encounters a proton, for example.

However, the creation of such pairs of real particles is not the only kind that occurs. According to the uncertainty principle, pairs of what are called virtual particles can pop into existence for extremely short periods of time. Since energy and time are related in the same way as position and momentum, energy uncertainties become quite large when one deals with very short periods of time. So suppose we have a situation where there is sometimes not enough energy available to create a particle-antiparticle pair. No matter. For short times, such pairs can be created from energy uncertainties instead. For example, when one deals with a period of time as short as 10^{-21} seconds, the energy uncertainty becomes so great that an electron-positron pair can be created. Similarly, the much heavier proton-antiproton pair can exist for a span of about 10^{-24} seconds.

Such virtual particles do not go their separate ways after their creation. Instead they undergo mutual annihilation almost immediately. It is as though energy they have "borrowed" must be paid back before anyone notices that it is gone.

As a result, virtual particles have never been observed directly, and most likely they never will be. However, they are quite real, and their brief existence makes itself felt. For example, the creation and annihilation of virtual particles inside an atom causes slight shifts in the frequency of the light that the atom emits when it makes a transition from one energy state to another. These shifts have been measured, and the predictions of theory have been verified to a precision of about one part in 10 billion. Since this kind of accuracy is rare, one can only conclude that the existence of virtual particles is one of the best-confirmed ideas in physics.

QUANTUM FLUCTUATIONS

According to the uncertainty principle, pairs of particles and antiparticles are constantly being created and destroyed everywhere. Such quantum fluctuations even take place in otherwise empty space. According to quantum mechanics, then, there is no such thing as "empty" space. On the contrary, even the most perfect vacuum is a sea of microscopic activity. (See Figure 8.)

Although the theory of virtual particles seems to work perfectly in the laboratory, it has created one of the outstanding unsolved problems of theoretical cosmology. If such quantum fluctuations are constantly taking place, then it seems to follow that any vacuum must contain a great deal of energy. But, according to general relativity, the presence of so much energy would overwhelm the gravitational effects of all the matter in the universe. In relativity, after all, matter and energy are equivalent; energy has gravitational effects too.

In fact, the discrepancy between theory and observation is enormous. If the energy density of empty space is calculated in this manner, one finds that the universe should have curled up into a tiny ball many orders of magnitude smaller than an atomic nucleus shortly after it was created. This would be equivalent to having a cosmological constant 10^{120} times larger than empirical data say it could possibly be.

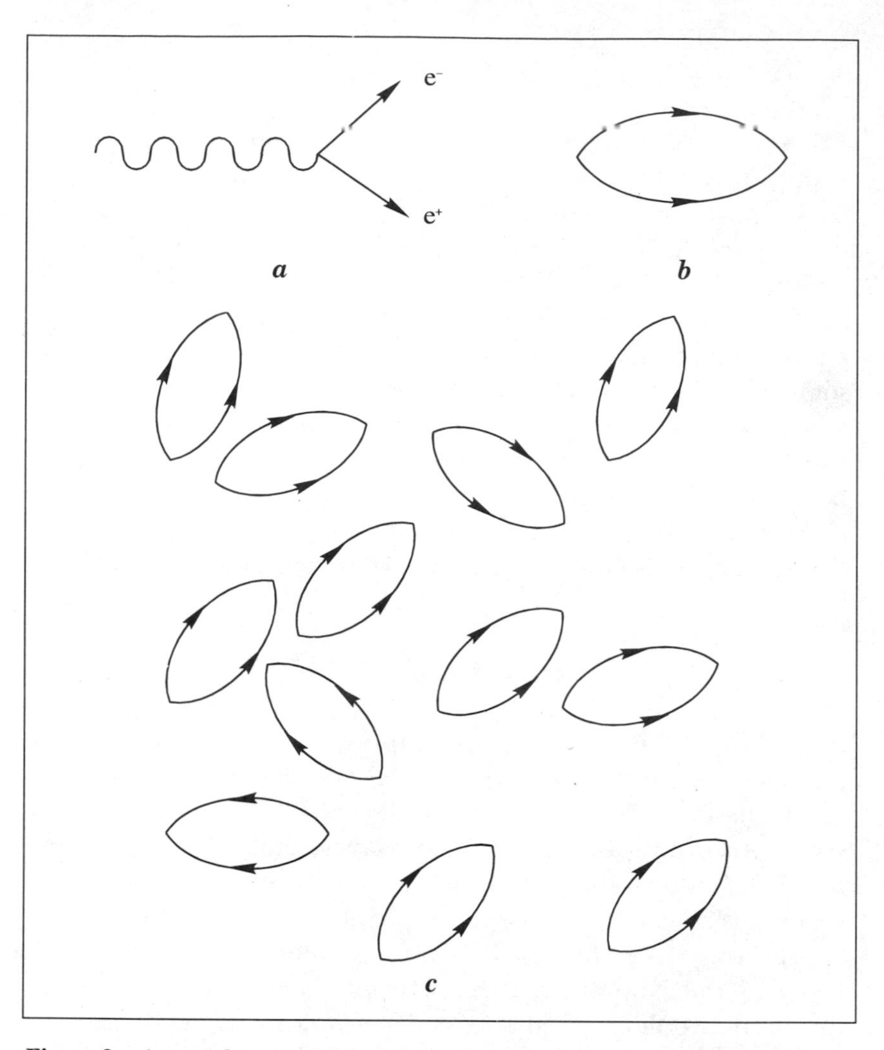

Figure 8 *A particle-antiparticle pair can be created out of energy* **(a)**. *Here, an electron (labeled* e^-*) and a positron (labeled* e^+*) are formed from the energy present in a gamma ray (wavy line). In diagram* **(b)**, *a pair of virtual particles are created from nothing. Since the energy that would be needed to give the particles a prolonged existence was not available, they annihilate one another shortly after they are formed. One often sees diagrams like this drawn with one of the arrows pointing in the opposite direction. This is not an indication of direction of motion, but a convention used to indicate that one of the particles is an antiparticle. I have used arrows pointing in the same direction for greater clarity. Diagram* **(c)** *represents some of the creation and annihilation events that are constantly taking place in "empty" space. For simplicity, all of the loops are depicted to be the same size. In reality, virtual particle pairs would exist for varying periods of time before they underwent mutual annihilation.*

The existence of this discrepancy does not necessarily imply that there is something wrong with the theory of virtual particles. It is more likely that there is some undiscovered flaw in the theories from which the constant is calculated, or there may be some unknown physical process at work which caused the cosmological constant to be canceled out.

Such cancellations do sometimes take place. For example, the universe in which we live seems to be electrically neutral. As far as scientists can tell, it contains an enormous number of positively charged particles and an enormous number of negatively charged particles, and the two numbers turn out to be exactly equal. If this were not the case, it would be possible to observe long-range electrical forces in the universe, and these would have very significant effects. After all, electromagnetism is a force which is many orders of magnitude stronger than gravity. The disparity between the two forces is illustrated every time someone picks up a piece of iron or steel with a magnet. The force that the magnet exerts on the object is stronger than the gravitational force exerted on it by the entire earth.

IS THE UNIVERSE A 90-POUND WEAKLING?

Another pair of quantities that seem to cancel one another out are matter and energy. If one were to add up all the matter in the universe, the result would be a very large number. As we shall see, most of the energy in the universe is gravitational, and if all this energy were added up, the result would be a very large *negative* number, for reasons that will be explained shortly. And, as far as scientists can tell, these two numbers are about equal: They more or less cancel each other.

The concept of negative energy is not nearly so abstruse an idea as one might think. With some further explanation, it is easy to see that the gravitational energy present in a system cannot be anything but a negative quantity.

What is gravitational energy? And when would the gravitational energy of two bodies be equal to zero? These questions can best be answered by means of some examples. Whatever gravitational energy is, it must be zero when bodies are so far apart that they experience no discernible gravitational attraction. The next question is, When the two bodies are very near to one another, do they have more energy or less? The energy must be less than zero because an expenditure of energy would be required to force them apart. In other words, as more and more energy is put into the system, the farther apart the two bodies are. When a great deal of energy has been expended, the total gravitational energy will rise to zero.

One can also look at the same process in reverse. Suppose that a comet is so far away from our sun that the gravitational attraction is nil. The gravitational energy must therefore be close to zero. Now suppose that the comet approaches our sun. As it does, the sun's gravitational pull will cause it to travel faster and faster: Its energy will increase. According to a law of physics called the law of conservation of energy, energy can be neither created nor destroyed; it can only be transformed into another kind of energy, or into matter. Energy is changed into matter, for example, when a particle-antiparticle pair is created. The reverse process takes place then matter is transformed into energy in the nuclear reactions that take place within the sun, or in the explosion of a thermonuclear bomb. The transformation of one kind of energy into another is an even more familiar process. An automobile engine converts the chemical energy present in gasoline into energy of motion. Similarly, hydroelectric power involves the conversion of the energy present in falling water into electricity. This electrical energy can, in turn, be converted into heat energy by an electric range, or into energy of motion in any of a number of different household appliances, or into light energy in an incandescent bulb.

Since a comet is not an object in which nuclear reactions take place, it is safe to assume that no energy is being created or destroyed when it enters the solar system. However, as the comet's orbit brings it closer to the sun, it begins to move faster and faster. In other words, its energy of motion is increasing. But this energy has to come from somewhere, and there can only be one source. As the comet's energy of motion becomes greater, its gravitational energy must decrease. But since it began with essentially zero gravitational energy when it was very far away from the sun, this quantity can only be negative when it is inside the solar system.

Energy exists in many different forms. However, when one considers the universe as a whole, gravitational energy is by far the most important. It exists in such huge quantities that all other forms are insignificant by comparison. The reason is that gravity is the only force that operates over long distances. Galaxies exert gravitational forces on other galaxies, and clusters of galaxies can feel one another's presence even when they are hundreds of millions of light years apart. When one considers the energy-matter balance in the universe, gravitational energy is the only form of energy that has to be taken into account.

To summarize: When one estimates the quantity of matter in the universe, a very large number is obtained. And when one estimates the amount of energy in the universe, one obtains another very large number, but one that is negative. If these two numbers are added together (something which Einstein's equation $E = mc^2$ allows us to do), one gets a result that *might* be zero. No one knows whether it is exactly zero. Neither matter nor energy can be measured with great enough precision to allow one to form any firm conclusions. However, the fact that the sum of the two might be zero has significant implications.

WAS THE UNIVERSE CREATED OUT OF NOTHING?

It would be possible to say that theories of nothingness have become an important part of cosmology in the last few years. There now exists a new field, known as quantum cosmology, which examines the ways in which the universe might have come into existence. And more often than not, scientists in this field form hypotheses about ways in which it might have been formed out of nothing.

Quantum cosmology is not as quirky a field as one might think. Trying to imagine a universe being created out of something would be considerably more bizarre. In fact, a theory which attempted to do this would not explain very much at all, for one would then have to answer the question of where that primordial "something" came from.

One cannot travel 10 or 20 billion years back in time to observe how the universe began; however, it is possible to speculate about the conditions that might have existed before the beginning of the inflationary expansion. Indeed, it is possible to go back even further and speculate about the moment of creation itself. And if this can be done, there is no reason, say quantum cosmologists, why attempts should not be made to find connections between conditions in the very early universe and the structure of the universe today. In other words, if scientists can create a plausible description of the manner in which the universe was created, they should be able to draw conclusions about what it should look like now. And if they can do that, it should be possible to work the other way around. If our universe does turn out to have the predicted appearance, then this would confirm the assumptions about its creation.

Reliance on this logic is central to quantum cosmology. Among the pioneers in this field are physicists James B. Hartle of the University of California at Santa Barbara, Alexander Vilenkin of Tufts University, Andrei D. Linde of the Lebedev Physical Institute in Moscow, and Stephen Hawking. These scientists have suggested a number of different plausible ideas about the origin of the universe. At the moment, there is no way of knowing which, if any, of these ideas is most likely to be correct, but it is well worth our while to take them seriously.

WHAT TO DO WHEN GENERAL RELATIVITY NO LONGER WORKS

At first, it may seem a little odd that such a field as quantum cosmology should exist at all. After all, quantum mechanics is the field of physics that deals with the behavior of matter on the submicroscopic level. Scientists generally use Einstein's general theory of relativity to describe the structure of the universe as a whole. So why do scientists not apply it to conditions in the preinflationary universe also?

They would if they could. Unfortunately, no theories have universal validity, and general relativity is not an exception. Einstein's theory breaks down when one considers events that took place at times very close to the beginning. Conditions then were so extreme and gravitational fields so intense that his theory would have to be replaced by something else if conditions were to be described accurately.

The first signs of trouble appeared during the 1960s, when Hawking and British mathematician Roger Penrose proved a series of theorems which proved that if one tried to use general relativity to look at times all the way back to the beginning, one would get infinite quantities. They found that, if general relativity was totally correct, the universe had to have begun in a state where all the matter that it contained was compressed into a mathematical point called a singularity. In a singularity, the density of matter and the curvature of space would have to have been infinite. After all, if everything is compressed into zero volume, infinities cannot be avoided.

Penrose and Hawking were not saying that a singularity actually existed at time zero. Rather, their theorems seemed to indicate that general relativity broke down under extreme conditions, such as those that existed at the first moments of time. When infinite quantities appear in a theory, that is always a sign that something has gone wrong. After all, true infinities are not encountered in nature.

There was nothing very surprising about the Hawking-Penrose results. Scientists had known for some time that the two major theories of modern physics, quantum mechanics and general relativity, were inconsistent with one another. They already knew that general relativity would have to be replaced by a quantum theory of gravity if one were to have any hope of describing the action of gravitational forces under conditions where quantum effects became important.

Before I go on, I must digress a bit and mention that, nowadays, there is some hope that the concept of superstrings may lead scientists to such a theory. A successful superstring theory would be a quantum theory of gravity, as would any theory that unified all four forces. If such a theory were found, both gravity and the forces on the submicroscopic level could be described within a single theoretical framework.

Unfortunately, superstrings are mathematically very complicated objects, and it is not likely that scientists will be able to develop a workable theory incorporating them until some time in the next century, and they may not, in the end, succeed. Some eminent scientists have suggested that the quest for a usable theory could be a blind alley.

Thus, at present, there is only one theory that is of use to scientists who want to speculate about the beginning of time, and that is quantum mechanics. It may seem a little strange that a theory of the behavior of submicroscopic objects should be the one used to describe the origin of that very large object we call the universe. However, there is no choice. There is nothing else that works. And if one really believes that quantum mechanics—which may be the most successful theory of physics ever devised—is applicable under all circumstances, there is no reason not to forge ahead.

IMAGINARY TIME

Although there are a number of different hypotheses concerning the origin of the universe, one is far better known than any of the others: The no-boundary proposal. This theory uses a concept known as imaginary time. It was devised by Hartle and Hawking and is described in Hawking's best-selling book, *A Brief History of Time.*

The first thing that should be pointed out is that the term *imaginary* should not be taken literally. When Hartle and Hawking use it, they are not speaking of ghosts and elves, but of something that is quite real. Hartle and Hawking are using the word imaginary in a technical, mathematical sense. They are suggesting that time may lose its ordinary, timelike character as one extrapolates to points close to the origin of the universe. In their theory, time becomes something resembling a spatial dimension at very early "times." Thus the universe has no real beginning for the simple reason that, if one goes back far enough, there are no longer three dimensions of space and one of time, but only four spacelike dimensions. Since the universe is assumed to be closed in this theory, it follows that it has no boundaries either in space or in time. A closed universe, one should recall, has no spatial boundaries.

It is easier to visualize such a situation than one might think. If general relativity held, the classical universe could be considered analogous to a cone, with the current universe represented by the fat end of the cone and the universe at its beginning represented by the point of the cone. As one goes back in time, the diameter of the cone (analogous to the diameter of the universe) gets smaller and smaller until all three spatial dimensions shrink to zero at the singularity. In the Hartle-Hawking theory, on the other hand, the cone would have a rounded cap. Time does not "keep on going," but instead becomes something other than time when one projects back into the past. Instead it cooperates with the three spatial dimensions to create a four-dimensional "sphere."

It can be said, then, that the Hartle-Hawking universe has no beginning, and just as it has no beginning, it has no end. At some time far in the future, according to this conception, the universe will be contracting toward a singularity (the big crunch). But it will never get there. Instead, time will become imaginary again, and there will be another rounded cap. (See Figure 9.)

Does this theory necessarily imply that the universe must be closed? According to Hartle, it doesn't. He points out that quantum cosmologists generally work with the conception of a closed universe because it is simpler. There are fewer mathematical complications when one is dealing with a finite object. However, there is no reason why the same methods could not be applied to an infinite, open universe also.

SOMETHING OUT OF NOTHING

If the universe had no beginning and no end, the problem of why it should have been created at a particular moment in time loses meaning, because there is a point at which time ceases to exist. However, the no-boundary proposal is by no means the only possibility, and so it is not certain that we have to avoid the problem of figuring out exactly what that first moment of time was. For example, it is conceivable that the universe began as a kind of quantum fluctuation. If it did, the question of why this fluctuation should have happened at a particular "time" is again meaningless. After all, quantum fluctuations are random events. (Here, I am ignoring the question of whether there even was any such thing as time before the creation of the universe. It is possible that both space and time were created together when the universe sprang into existence.)

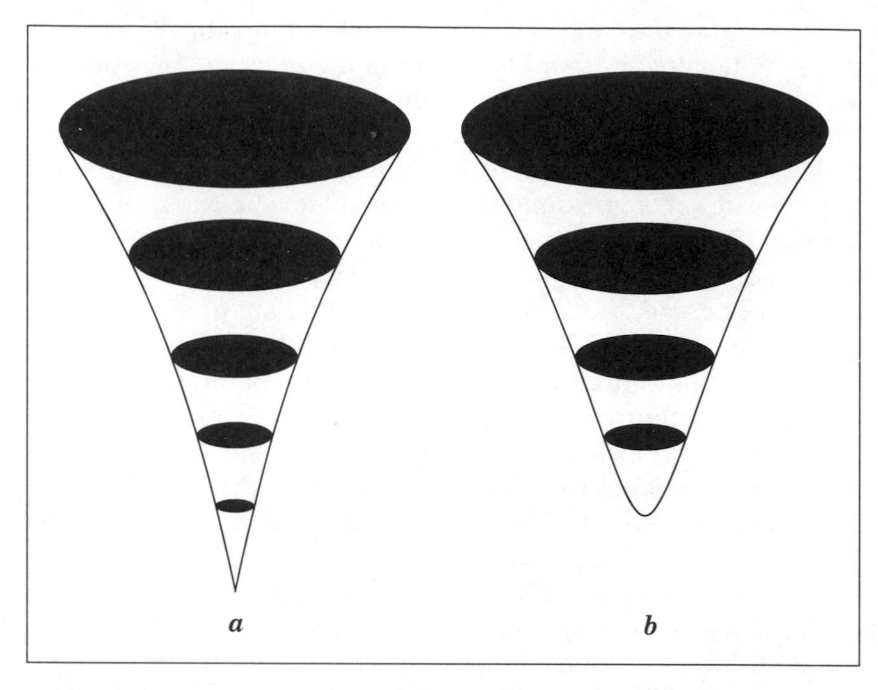

Figure 9 *In diagrams **(a)** and **(b)**, the expansion of the universe is represented by a series of dark circles. Each circle represents a stage in the universe's evolution. In diagram **(a)**, all the matter in the universe is compressed into a mathematical point at time zero. In such a state, called a "singularity," the density of matter and energy would become infinite. In the theory proposed by Hartle and Hawking **(b)**, there is no singularity because time does not extend all the way back to this zero point. Instead, time curves in upon itself and becomes something resembling a spatial dimension. The origin of the universe is thus represented not by a point, but by a kind of rounded cap.*

The idea that the universe might have begun this way is not a new one. It was first suggested as long ago as 1973 by the American physicist Edward Tryon, who pointed out that if the mass-energy content of the universe was indeed zero, then Heisenberg's uncertainty principle would allow it to be created out of nothing, and to exist for an indefinite period of time.

Tyron's proposal wasn't a fully worked out theory, but it apparently sounded like a good idea, for other physicists soon began to elaborate upon it. For example, in 1978, four Belgian physicists, R. Brout, P. Englert, E. Gunzig, and P. Spindel (I apologize for giving only their initials, but this is the way they put their names on the paper in which they published their results) suggested that the universe might have begun with the creation of a single particle-antiparticle pair. And, in 1981, physicists Heinz Pagels and David Atkatz of Rockefeller University proposed that the universe might have begun, not with the creation of particles, but with a sudden change in the dimensionality of space. According to their hypothesis, the universe initially contained no matter and had a large number of spatial dimensions. The big bang could have taken place, they suggested, when the universe "crystallized" into its present form.

The idea of creation out of nothing has more recently been worked out in great detail by Alexander Vilenken. In Vilenken's theory, the "quantum fuzz" from which the universe emerged didn't even have a definite dimensionality. According to his theory, the very concepts of space and time have meaning only after the universe comes into existence.

SELF-REPRODUCING UNIVERSES

Andrei Linde takes the idea of creation out of nothing even further. If one universe can come into existence in this manner, he asks, why can't many of them? According to Linde's chaotic inflationary universe theory, new universes are being created all the time through a kind of "budding" process. Tiny balls of spacetime called "baby universes" are created in a universe like our own. These baby universes then undergo inflationary expansions, and many of them evolve into universes resembling the one in which we live.

This theory does not imply that space travelers are likely to encounter full-fledged universes blocking their path if they ever venture out toward the stars. The universes that budded off from our own would most likely be no more observable than virtual particles. In fact, there is good reason to expect that they might "pinch off" from our spacetime and then disappear. For a brief moment, a thin strand of spacetime called a wormhole might connect parent and baby universes. But most likely this wormhole would quickly vanish.

In Linde's theory, the number of possible universes would be limitless. The universes that bud off from a parent universe could then have offspring of their own, and these could in turn give birth to numerous others. Naturally there is no reason to think that all these countless universes would be alike. Most likely, they would not be. In some of them, inflationary expansions might never end. Others might pop into existence, expand for a short time, and then collapse into nothingness again. As they did, they might be reabsorbed into their parent universes. Alternatively, the umbilical-cord-like wormhole might already have been broken off. There is no reason to think that wormhole connections between universes would have to be permanent. Most likely, they would not be.

HAWKING'S WORMHOLES

It is probable that, if other universes exist, we will never be able to observe them. However, their existence could nevertheless have real physical effects, just as the creation and annihilation of unobservable virtual particles give rise to physical phenomena that can be measured in the laboratory.

For example, Hawking has suggested that subatomic particles may constantly be traveling through wormholes from one universe to another. According to Hawking's theory, this would not be an observable phenomenon. The wormholes, which would presumably have diameters about 10^{20} times smaller than the dimensions of an atomic nucleus, could never be detected—nor would they exist for very long. A typical wormhole would most likely remain in existence only for about 10^{-43} seconds.

Even though wormholes were so small and so short-lived, however, subatomic particles could still pass through them. But one could not observe this process either. According to Hawking's hypothesis, every time that a particle left our universe, an identical particle would come into our universe to replace it. And since (for example) every electron looks exactly like every other electron, we wouldn't be able to tell that anything had happened.

But, according to Hawking, such particle exchanges could have real physical effects. His theory seems to imply that particles that can disappear into wormholes will appear to have greater masses than particles that always remain in the same universe. In

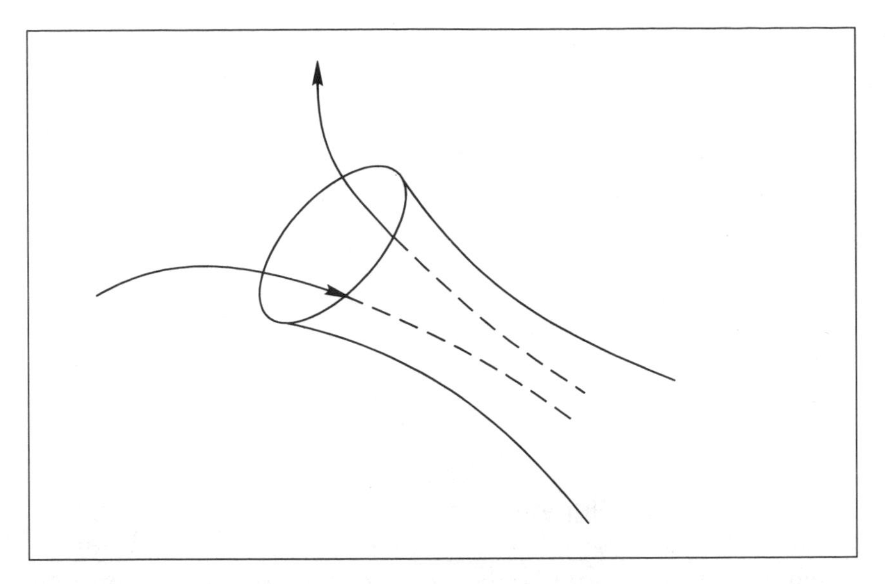

Figure 10 *According to a hypothesis proposed by the British physicist Stephen Hawking, subatomic particles may be constantly disappearing into submicroscopic wormholes and traveling to other universes. In this diagram, a particle enters a wormhole while an identical particle emerges from it and enters our universe. Since such wormholes would be too small to be seen, such events would be unobservable. However, according to Hawking, particles might acquire mass by just such a process.*

fact, a particle that never went through a wormhole might not have any mass at all. (See Figure 10.)

If Hawking's theory could be refined to the point that it was able to predict the masses of electrons, protons, and other particles, then we would have to take the idea of wormhole exchange quite seriously. We might never be able to observe these other universes. However, it is not always necessary to be able to see something to be convinced that it is real. No one has ever seen a magnetic field either, yet few people doubt that magnetism exists.

No one has yet been able to use Hawking's theory to produce reasonable-looking predictions, so perhaps the idea must be considered quite speculative. One should not label it idle speculation, however. Some of the greatest advances in the history of physics have taken place because scientists who were trying to solve some problem proposed one speculative idea after another. If one "crazy" idea didn't work, they kept on proposing such ideas until they found the correct one.

INITIAL CONDITIONS

If we are told that someone has thrown a ball into the air, it is not enough to know the law of gravity if one is going to predict where it is going to land. It is also necessary to know precisely how hard the ball has been thrown, and in what direction. If an astronomer observes a comet entering the solar system, he cannot tell at once how close it will pass to the sun, or when its nearest approach will take place. He must first obtain information about the comet's exact position at some moment in time, and determine its velocity and the precise direction in which it is moving.

Scientists call such quantities initial conditions. In every field of physics, initial conditions have to be known before the future behavior of a body or system of bodies can be predicted. And if initial conditions cannot be determined, then the statements that can be made are not particularly useful. Without a knowledge of initial conditions, one is reduced to saying things like "A ball that is thrown up must come down," or "Comets often enter the solar system, and then, sometime later, they leave again."

Initial conditions don't have to be those that existed at the earliest possible moment. It is often more useful to know something about the behavior of an object at some later time. If we know what properties the ball had immediately after it was manufactured, for example, that will hardly allow us to calculate where it will land after being thrown into the air at some later date.

If one can speak of initial conditions for a ball or for a comet, then there seems to be no reason why one cannot speak of the initial conditions of our universe. Possibly this may even be a more clear-cut concept. Our universe, after all, is a self-contained system, and it is not necessary to worry about the initial conditions of other objects that might affect its behavior.

Moreover, when one speaks of the initial conditions of our universe, it is not necessary to speak of the very first instant of time. As we have seen, there may not even be any such thing as a first instant. If the no-boundary proposal is correct, for example, time is initially spacelike.

And, in Vilenken's theory, the universe emerges from a state in which time as we know it did not exist.

Quantum cosmologists do not attempt to say how hot or how dense the universe was at some particular moment. They do not try to guess how many particles of what varieties existed. They do not try to imagine what dark matter might be. Yet they do try to construct plausible initial conditions. Then they try to see how the universe might have evolved thereafter.

QUANTUM COSMOLOGY AND PREDICTION

The theories of quantum cosmology are very speculative. However, the quantum cosmologists are not simply spinning fantasies about the origin of the universe. They are engaged in quite a serious endeavor.

In principle, there is no reason why a hypothesis about the initial conditions of the universe could not be used to make predictions about the structure of the universe today. If these predictions were confirmed by astronomical observation, then scientists would have good reason for thinking that they had discovered how the universe began. And that would be an achievement indeed.

In fact, this has already happened to some extent. For example, the no-boundary proposal predicts that our universe should be one that is spatially flat, homogeneous (having about the same mass and energy density everywhere), and isotropic (having about the same appearance in every direction). It also predicts that we should live in a universe that does not exhibit large-scale quantum fluctuations.

One can't say that the confirmation of these predictions really confirms the no-boundary proposal but only that they don't falsify it. After all, flatness, homogeneity and isotropy are predictions of the inflationary universe theory too. And the lack of large-scale fluctuations is a result of the small size of Planck's constant. There could be many kinds of universes other than one beginning in imaginary time in which Heisenberg's uncertainty principle operated only on the submicroscopic level.

However, the fact that quantum cosmology can yield predictions that are fulfilled is an indication that a reasonable beginning has been made. And so there is every reason to hope that more substantial results may be obtained in the future. In fact, quantum cosmology may eventually be able to tell us why the laws of physics have the form they do. There is some reason to think that, if the initial conditions of the universe had been different, then the laws of nature would have been also.

For example, scientists have long wondered why the electron should have a certain specific mass. Why shouldn't it be half that heavy? Or ten times heavier? Why should the proton have a mass that is 1837 times as great as that of the electron? Why should the basic unit of electrical charge have a particular value? Why should there be four basic forces, and not three or five? Why should the force of gravity have a particular strength? Why shouldn't it be stronger, or weaker? Why should there be exactly three different kinds of neutrino? Why should our universe have three directions of space, and not some other number? Why doesn't it have a hundred, or even a thousand, different spatial dimensions? Why is the cosmological constant either zero, or so close to zero that it cannot be measured? Why is time so homogeneous? Why couldn't it run at different rates in different parts of the universe?

No one knows the answers to such questions. Many physicists hope that superstring theory, which attempts to combine all four forces of nature in a single theoretical framework, may eventually provide many or all of the answers. But it is also possible that it may not. At this moment, no one knows where superstring theory will lead. Indeed, there is a possibility that this particular theoretical endeavor may not be successful at all.

If superstring theory does not provide answers to these questions, quantum cosmology may. It is not unreasonable to think that some or all of the laws of nature may have been determined by the initial conditions of our universe. And it is also conceivable that quantum cosmology will eventually answer certain important but less fundamental questions, such as those concerning the nature of dark matter and the mechanisms that led to the formation of galaxies.

8

What Is Time?

I have previously discussed time in a number of different contexts. But there is much more that can be said about this complex and baffling topic, and there are numerous questions that can be asked. For example: How do we know that time proceeds in the same direction everywhere in the universe? Why couldn't there be places where it goes backward? For that matter, if time runs forward in an expanding universe, why shouldn't it run in the other direction in a universe that is contracting? And if it does run backward in a collapsing universe, perhaps our universe does not have a beginning and an end, but rather two beginnings.

These questions are not as crazy as one might think. Though there is no evidence that time ever runs backward, there is nothing in the laws of physics that says that it can't. Though the great majority of physicists are reasonably certain that time would continue to behave in a normal manner in a contracting universe, certain eminent scientists, such as the American physicist Thomas Gold and, later, Stephen Hawking, have speculated about the possible existence of backward time in a contracting universe.

There is much about the phenomenon of time that scientists do not yet understand. In particular, there are many aspects of the directionality of time that remain mysterious. Scientists have been trying to solve the problems associated with time for more than a century. The more they have learned, the more baffling the question has become.

TIME-REVERSIBLE LAWS

The problem is that all of the basic laws of physics are time-reversible. In other words, they would all remain perfectly valid if time ran in the opposite direction. For example, suppose one could somehow make a videotape of the planets of the solar system as they revolved around the sun. If the tape were then run backwards so that the planets were moving in the opposite direction, nothing would be seen that would contradict any physical law. The motion of the planets would still be described by Newton's (or by Einstein's) law of gravitation. And yet, running the tape backwards would be equivalent to viewing the planets as they moved back in time.

The same considerations apply to events that take place on the subatomic level: They too are time-reversible. Suppose, for example, that an atomic nucleus emits an electron. This is a common kind of radioactive decay, one that happens all the time. But the same process can take place in reverse: It is also possible for a nucleus to absorb an electron. Thus if one were to videotape a radioactive decay and then run the tape in reverse, nothing would be seen that would contradict any of the basic laws of physics.

THE STATISTICAL CHARACTER OF TIME

Nothing is more obvious than the fact that time does have a direction. We remember yesterday and we anticipate tomorrow. The reverse never happens. We are born. We grow old. We die. We never live our lives from the end to the beginning. The inanimate objects that fill the world around us exhibit irreversible behavior too. If a wine glass is dropped on the kitchen floor, it will shatter. The pieces of a broken glass never come together to make the glass whole, and they never fly upward into one's hand. If you or I watched a videotape and saw something like this happening, then we would know that the tape was being run backward. Similarly, when an ice cube is placed in a glass of water, it melts. And the embers of a dying fire always fade away; they never grow brighter, burst into flame, and reconstitute themselves into an unburnt log.

And yet the laws of physics do not forbid the time-reversed versions of the events we observe. According to these laws, it is not impossible that the pieces of a broken glass should spontaneously come together, or that an ice cube could form in a glass of cold water. These events are only very improbable. The reason that we do not observe them is not that they could not happen, but that they are so unlikely that we can safely disregard the possibility.

The reason for this statement is that the individual molecule or atom or subatomic particle knows nothing of the direction of time. It is only very large numbers of particles that are time-dependent. An individual air molecule may be anywhere in a room, moving in any direction. But there are trillions of trillions of molecules in an average-sized room, and the laws of probability ensure that these molecules will be evenly distributed. Therefore, it is theoretically possible that all of the air molecules could suddenly decide to collect in one corner, causing everyone in the room to die of asphyxiation. But such an event is as improbable as that of a broken glass coming together. It is entirely reasonable to assume that it will never happen.

THERMODYNAMICS

The field of physics that describes the average, statistical behavior of matter is called thermodynamics. In thermodynamics, certain concepts are introduced which cannot be applied to the behavior of individual particles but which are essential when large ensembles of particles are considered. One such concept is that of temperature. One can speak of the velocity or energy of an individual particle, but never of its "temperature": Temperature is a measurement of the average energy of a large number of particles. The greater this average is, the "hotter" an object is said to be.

Another concept that is quite important in the field of thermodynamics is that of entropy. Entropy is often equated with disorder, and it is said that, as the entropy of a system increases, the order in that system diminishes. For example, as an ice cube melts, entropy increases. The ordered, crystalline structure of ice is replaced by the more random arrangment of molecules of water.

Although this is quite true, there are also other ways to explain the concept. I prefer to say that, as the entropy of a system increases, its various components come into equilibrium with one another. A low-entropy state would thus be one in which there was a great deal of disequilibrium.

Although *disequilibrium* is a somewhat awkward-sounding word, it describes a concept that is quite simple. When water falls from one level to another, it can do work by moving a waterwheel or the blades of a turbine. At one time, millers used this energy to grind flour, and today we use it to generate electric power. This can be done, not because the water contains energy, but because there are energy differences. If a lake is situated at an altitude of one mile, for example, the water that it contains will have a great deal of gravitational energy compared with water at sea level. But if the lake has no outlet, this energy cannot be used. On the other hand, a lake at an altitude of one thousand feet possesses much less gravitational energy. However, if the water can be made to fall to a lower level, this energy can easily be used.

Similarly, the oceans contain a great deal of heat energy. After all, there is energy in any substance that is not at a temperature of absolute zero, and the oceans contain a great deal of water. However, no one has ever made use of all this heat energy to power a ship, and no one ever will. On the other hand, the total quantity of energy in a hot boiler is relatively small. Even though it may be hotter, it contains much less water than an ocean. However, a steam engine can produce a great deal of power because it allows energy to flow from one level (the hot boiler) to another (the cool surroundings).

Thus it is differences of energy, or disequilibrium, that are important, not the absolute quantity of energy in a system. This provides the motivation for inventing a property that will measure disequilibrium, or the lack of it. This concept is called entropy. When there is a great deal of disequilibrium in a system—when energy differences are pronounced—the entropy of that system is said to be low. When these energy differences disappear, so that no useful work can be done, entropy is said to be high. If one prefers to think of entropy in terms of order and disorder, the parallels are obvious. A low-entropy system is one that possesses discernible structure. In a high-entropy system, things tend to even out; energy differences disappear.

THE SECOND LAW OF THERMODYNAMICS

I suppose that, by now, most people have heard of the second law of thermodynamics.* According to this well-known principle, the entropy of any isolated system must always increase. The energy differences, or disequilibrium, of the system will tend to dissipate, and the ability of the system to do useful work will diminish, even though the total energy content may remain the same.

Our solar system is really not an isolated system. Energy from our sun radiates out into space, whereas an isolated system would not interact with its environment at all. Thus the second law of thermodynamics technically cannot apply here. Nevertheless, the solar system can be used as a rough example of increasing entropy. At present, life on earth is maintained by the flow of energy from the sun. The sun, after all, is very hot, and the earth is relatively cool. However, billions of years in the future, when the sun is a cool white dwarf, the flow of energy to whatever planets still exist will be greatly diminished, and the entropy of the system will be high. It is not likely that this high-entropy solar system will be capable of providing environments that are very hospitable to life. Once again, it should be noted that the important thing here is the existence of differences in energy levels, not total energy. Three billion years from now, the sun will possess much less total energy than it does now, but this

* Naturally there is also a first law. The first law of thermodynamics is nothing more than a restatement of the law of conservation of energy.

will not make much difference to the existence of life on earth (if life still exists) because energy will still be flowing from a hot place to a cool one.

THE ARROWS OF TIME

The second law of thermodynamics can be made into a definition of the direction of time. In other words, the direction of increasing entropy can be regarded as an "arrow" of time that always points toward the future. Whereas the more fundamental laws of physics are time-reversible, the second law of thermodynamics is not. Entropy is one physical principle that distinguishes between past and future.

The expansion of the universe provides us with another arrow of time. The future can be defined as the direction in which the size of the universe increases. It is this fact which raised the question of what would happen in a contracting universe. Though there is no particular reason to suppose that the direction of time would actually reverse, some physicists have entertained this as a possibility. But perhaps it is significant that Hawking now concludes that there would not be any reversal of the direction of time in a contracting universe.

A third arrow of time is that of human consciousness. As I have pointed out previously, we remember the past and anticipate the future.

A fourth is the electrodynamic arrow of time. Electrodynamics is the field of physics that deals with moving electrical charges and with such phenomena as electromagnetic radiation. Like other basic laws, those of electrodynamics are time-reversible. Since they don't distinguish between the two directions of time, they predict that light and other forms of electromagnetic radiation should travel into the past as well as into the future.

But time-reversed waves that travel into the past are not observed, and it is not entirely clear why they are not. If the laws of electrodynamics are taken at face value, one has to conclude that astronomers should be able to look into the future (by looking at light coming from the future) as easily as they look into the past. Similarly, if one went far enough out into space, one could presumably detect television broadcasts from next week (one would have to go out into space to see last week's television too).

There does exist a theoretical argument, developed by the American physicists Richard Feynman and John Archibald Wheeler, which purports to show that time-reversed radiation will cancel out and that only the familiar kind of waves that travel into the future will remain. But their argument is complicated, and it seems to work only if the universe happens to be closed.

Alternatively, it may be that time-reversed radiation is somehow forbidden by the second law of thermodynamics. Even in this case, radiation that traveled into the past would not be an impossible phenomenon. It would only be very improbable. However, the connection between the thermodynamic and electrodynamic arrows of time is not very clear.

To make matters worse, it appears that the thermodynamic arrow of time probably cannot be considered very fundamental. It is possible to conceive of universes where it does not exist at all. Quantum cosmologist James Hartle points out that the very existence of the thermodynamic arrow may be a consequence of the initial conditions of the universe. And this raises the possibility that there may be universes in which it does not exist. It is difficult to imagine what such universes would be like. Could time run in different directions in different parts of them? Could they contain regions in which time could not be defined at all?

There is also a fifth arrow of time, one that is related to the behavior of a subatomic particle called the kaon. But its existence hardly clears up any problems. Not only is it not known how this arrow is related to the other four, but it is not very clear why it exists at all.

The kaon provides an exception to the rule that all fundamental physical laws are time-reversible. On rare occasions, the kaon will decay into other subatomic particles in such a way that it would be possible to watch a videotape of the event and tell whether the tape was being run backwards or forwards. Most of the time, the kaon's decay is perfectly time-reversible. It decays in such a manner that the same event could easily take place in reverse. The anomalous decays—the ones that are not time-reversible—occur less than one percent of the time. But they occur.

To make matters worse, the existence of these decays seems to be of no particular importance. The kaon is not a constituent of ordinary matter, and it plays no role in the decay of atomic nuclei (it is another of those particles seen only in the laboratory).

The kaon's arrow of time is something that is observed only in the laboratory.

There is much more that can be said about the arrows of time. Unfortunately, it is impossible to go into the matter fully here, and I must refer the reader to the various books which deal with the subject (including my own book, *Time's Arrows*). However, this brief discussion should make it clear that time is not the simple thing it seems to be, and that it presents problems that scientists don't know how to begin to solve.

Could it be that time is not the fundamental quantity it seems to be? Could there be universes without time? No one knows how to approach those questions either. It would be difficult even to speculate about a universe without a time dimension. For example, in the field of quantum cosmology, one normally begins by assuming that the universe is something that has dimensions of space *and* time. Without a time dimension, one could hardly relate initial conditions to anything that happened later (because there would be no such thing as "later").

TIME TRAVEL

Perhaps the question "What is time?" is too far-reaching and too abstract. Perhaps one should approach the subject in a more modest manner. Instead of trying to determine what time is, perhaps one should try to understand some of its properties. For example, the idea that it might be possible to travel back and forth in time has been a staple of fiction ever since the British novelist H.G. Wells published *The Time Machine* in 1895. Is it conceivable that time travel might actually be possible?

Most physicists hope that it is not. If it were, all their ideas about causality would have to be thrown overboard. If the future could influence the past in some way, then physics would practically cease to exist as a science. After all, most physical laws are based on the idea that causal influences can only propagate from the past toward the future. When one flips a switch, the lamp goes on. When one strikes a billiard ball, it begins to move. Things just don't happen the other way around.

Travel into the past could produce paradoxes as well. For example, someone could go back in time and kill a parent before he or she was born. Even being able to send messages into the past would cause problems. For example, suppose my brother dies in a plane crash. In order to allow him to avoid this fate, I send a message to his past self warning him not to get on the plane. He follows my advice because he knows that I am a very trustworthy person. But because he does not get on the plane, he is not killed. But if he is not killed, I have no reason to send a message into the past to warn him about anything. Because he receives no warning, he does get on the plane and is killed. Therefore I do send a message into the past warning him . . . . This sort of thing can go on and on in an endless loop.

At the risk of straying somewhat from the subject, I would like to relate the plot of a story by the late science fiction author Robert Heinlein. In this tale, a rather inexperienced young woman is seduced by a man from the future. She becomes pregnant and gives birth to a child. She later undergoes a sex change operation and becomes a man. As a man she goes back in time and becomes her own seducer. But that isn't all. The child, a girl, to whom the woman gives birth is transported back in time too. She grows up to become the young woman. Time travel has created a situation in which mother, father, and child are all the same individual, who encounters him/herself while traveling back and forth in time.

INVENTING TIME MACHINES

Sometimes it seems that only people who possess a streak of the diabolical become physicists. This seems to be borne out by the frequency with which they have gone about inventing theoretical methods that would allow one to travel back in time. I can't help but suspect that one of the primary motivations for doing this is the desire to bedevil one's colleagues.

One of the first scientists to discuss time travel in a serious, technical manner was the Austrian-American mathematician Kurt Gödel. In a famous paper published in 1949, Gödel pointed out that some solutions to Einstein's general theory of relativity allowed travel into the past. At least this would be the case of the universe were rotating.

Fortunately (or unfortunately, if one is a member of the Devil's party), it appears that the universe is not rotating. We know this because if the universe exhibited any appreciable rotation, this would change the character of the cosmic microwave background. Since such rotation-induced effects are not observed, it is possible to conclude that, if there is any rotation, it must be very small.

Today scientists generally tend to think that Gödel's result was nothing more than a mathematical quirk, and that the rotation of the universe is exactly zero. But since it is not possible to prove this, in a certain sense the question is still open. Possibly quantum cosmology will eventually provide a solution to this problem. There may be some reason why a universe like ours cannot have any rotation. On the other hand, it is not inconceivable that there is a small loophole in the laws of physics that allows time travel to some limited extent.

TRAVERSABLE WORMHOLES

Einstein's special theory of relativity implies that no material object or causal influence can travel faster than the speed of light. It also tells us that if something could travel faster than light, then travel into the past would also be possible. For example, if a spaceship could travel at twice light speed, then it could go on a journey and return to earth before it left.

Similarly, if some object could travel instantaneously to some distant part of the universe, it could also travel into the past. After all, such instantaneous transportation would be nothing more than a special case of faster-than-light travel. And if it were possible to send instantaneous messages to distant locations, then messages could also be sent to past times.

Most physicists believe that such things are impossible. Unfortunately, they have to contend with the fact that general relativity contains yet another loophole that suggests a method by which such things might be possible. Perhaps, instead of "loophole," I should say "wormhole," for Einstein's theory suggests that wormholes might possibly be used for instantaneous communication between different parts of our universe.

I have previously discussed wormholes in connection with Linde's reproducing universe theory. There is no reason why wormholes—if such things really exist—would have to connect only different universes. A wormhole could just as well provide a bridge between two regions of our own.

If wormholes were microscopic and did nothing more than allow particle exchange of the sort envisaged by Hawking, they would be the source of no serious problems for time travel. After all, if a particle emerged from such a wormhole every time that one entered it, it wouldn't be possible to tell that anything had happened. Even if an individual electron might sometimes find itself journeying back in time, this event would basically leave the universe totally unchanged, with no untoward effects. It is difficult to see how one could even send messages into the past by such a method.

But if a macroscopic object would pass through a wormhole, that would be a different matter indeed. As one might suspect, this possibility has been a source of fascination for many physicists, and has been discussed in a number of different contexts.

The possible existence of traversable wormholes was first suggested years ago by theoretical physicists who were trying to understand the structure of black holes. The equations of general relativity seemed to imply that an object that entered a rotating black hole and that followed exactly the right trajectory once it got inside would not remain in the black hole: It would be transported through a wormhole to some distant region of the universe. As we have seen, this further implied the possibility of travel backwards in time.

No one was suggesting that someone actually try this. For one thing, the nearest black hole is light years away. For another, a journey into a black hole would undoubtedly be a one-way trip. If you or I entered one and discovered that things were not turning out as we had anticipated, there would be no turning back: We would be trapped inside the black hole forever. Furthermore, although such a journey might be theoretically possible, in practice a human being would not survive it. The enormous gravitational forces that exist near the surface of any normal-sized black hole would produce massive tidal effects that would rip any space vehicle apart and stretch any human being into a long, thin thread. To be sure, one could avoid experiencing such tidal forces by entering one of the supermassive black holes that presumably reside in the centers of galaxies. Though their masses are enormous, their larger size lessens the tidal effects near their surfaces. But I, for one, would not want to venture near one of these objects, because the radiation around them would simply be too intense. These are the black holes that presumably once powered quasars.

The theoretical possibility of travel through wormholes greatly bothered certain theoretical physicists, who were concerned with upholding the status quo. As it turned out, they needn't have worried. After further calculations were done to explore the possibility, the results indicated that though the existence of wormholes that could accommodate macroscopic particles was a theoretical possibility, travel through them was not. A wormhole that connected two regions of space might very well form, but if it did, it would close off so quickly that nothing could possibly get through it.

FISHING FOR WORMHOLES

Theoretical physicists, however, are nothing if not ingenious. It wasn't long before three of them thought of a way to bedevil the theoretical establishment again. In 1988, California Institute of Technology physicist Michael S. Morris (no relation to the author), Kip S. Thorne, and Ulvi Yurtsever published a paper in the journal *Physical Review Letters* in which they suggested that there was no reason why members of a highly advanced technological civilization might not be able to fish a microscopic wormhole out of the quantum soup in which it was formed and enlarge it to macroscopic dimensions. If such a macroscopic wormhole could be maintained, they said, it could not only provide a gateway to distant regions of the universe but could also function as a portal into the past.

Before I go on, I should probably repeat that no one knows whether such wormholes exist. General relativity seems to allow their existence, but there is nothing in the theory that requires them. Furthermore, if they have the very small dimensions that physicists generally ascribe to them—something of the order of 10^{-33} centimeters—then general relativity does not describe their behavior very well anyway, because they are so small. Only the long-sought-for theory of quantum gravity could do that.

Thus, though Morris, Thorne, and Yurtsever did make some very specific suggestions as to the kind of technology that might be used to enlarge such wormholes to macroscopic dimensions and to maintain them so that travel through them would be possible, one does not have to take the idea very seriously if one doesn't want to. It seems to be more science fiction than science. As a matter of fact, the method did make an appearance in a science fiction novel. When author Carl Sagan was writing his novel *Contact,* he asked one of the physicists to suggest a plausible method for faster-than-light travel, and duly incorporated the answer into his book.

TIME MACHINE II

The idea that traversable wormholes might be created seems far-fetched, to say the least, because it depends upon too many questionable assumptions. In particular, one must assume that some hypothetical civilization will somehow find a way to find, catch, enlarge, and maintain wormholes. Furthermore, any one of those four tasks could very well be impossible. And of course there is no evidence that wormholes connecting different parts of our universe exist in the first place.

Obviously, if we are to believe in the possibility of time travel, we need a better kind of time machine. In fact, one may well have been invented, in theory if not in reality. In March 1991, Princeton physicist J. Richard Gott published a paper (also in *Physical Review Letters*) in which he claimed to have shown that a pair of cosmic strings could open a route into the past.

As I have previously pointed out, a cosmic string—if it existed—would be a massive object with a density of trillions of tons per square inch. Therefore spacetime would be very strongly curved in the vicinity of the string because of the enormous strength of its gravitational pull. Furthermore, theory indicates, that if cosmic strings exist, they must travel through the universe at velocities approaching that of light. Under such circumstances they should be capable of producing some very dramatic effects.

When Gott looked at the equations of general relativity to see what they said about the curvature of spacetime in the vicinity of a cosmic string, he found that a cosmic string would, after all, offer no possibility of time travel. But when he began to consider what would happen if two cosmic strings happened to hurtle past one another in opposite directions, matters changed. He found that if a spaceship looped around the strings along a certain trajectory, it could travel into the past without exceeding the speed of light. In particular, it could follow a path that brought it back to its starting point in such a way that it would arrive before it had left.

This result created a certain amount of consternation in the theoretical physics community. Until then, it had not been necessary to take the traversable wormhole idea very seriously. But Gott's scenario for time travel was a different matter. Although cosmic strings were only theoretical objects, there was good reason for believing that they might exist. Furthermore, Gott's calculation had been relatively straightforward. He had apparently shown that, under the right conditions, general relativity did allow travel into the past. Even if this could not be accomplished in practice, it did appear to be theoretically possible.

In the eyes of many theoretical physicists, such a possibility should not exist in any rational universe, so a theoretical counterattack was launched almost immediately. Edward Fahri and Sean Carroll of the Harvard-Smithsonian Center for Astrophysics published a paper in which they claimed that Gott's time machine could be ruled out in an open universe. In such a universe, they said, it would not be possible to form such strings and to accelerate them to the necessary velocity.

The Fahri-Carroll analysis said nothing about closed universes, but at least it was a beginning. Shortly after it appeared, another thrust was launched. Theoretical physicists Stanley Deser of Brandeis University, Roman Jackiw of MIT, and Gerard 't Hooft of the Institute for Theoretical Physics in the Netherlands claimed to have found a fallacy in Gott's argument. When Gott disagreed, 't Hooft retorted by writing a paper in which he claimed to have found that time machines were impossible in a closed universe. If his argument was correct, and if the Farhi-Carroll analysis for open universes was also accurate, then time machines of this type could be ruled out completely.

Nevertheless, the controversy continued. Gott replied that his critics had made use of fallacious, circular arguments, and attempted to bolster his case by citing a paper by California Institute of Technology physicist Curt Cutler which, according to Gott, showed that time travel could be a real possibility. Then Stephen Hawking entered the fray. Time travel was clearly impossible, Hawking joked. If it were not, we would have to contend with hordes of tourists from the future. Hawking then backed up this argument with a more mathematical one in which he attempted to show that energy buildups would destroy a time travel loophole as soon as it was created.

THE CAFFÉ TRIESTE THEORY OF THE UNIVERSE

Most probably, then, time travel is impossible. But that will not prevent physicists from trying to construct plausible time travel scenarios. Even when these schemes turn out to contain flaws, something will be gained when they are subjected to intense theoretical scrutiny. After all, such controversy helps to throw light on the implications of the general theory of relativity, and on the kinds of spacetime structures that might be possible in our universe.

I frequently heard time travel discussed as I was researching this book. Every now and then, instead of studying a scientific paper in my office, I would go to a place called Caffé Trieste and read it over a cappucino or a glass of chianti or a beer. Caffé Trieste is an establishment in San Francisco's North Beach district that was founded by an Italian family in the 1950s, while the "beatniks" flourished. The cafe still retains its bohemian character, and it provides quite a congenial, romantic environment for the leisurely study of abstruse ideas.

One of the habitués of the Caffé Trieste is San Francisco physicist Jack Sarfatti, who has been trying to prove for years that quantum mechanics allows the possibility of signals from the future into the past. Most physicists would probably disagree with Sarfatti, and I suppose that I must include myself among them. Nevertheless, one of his most recent ideas is so intriguing that it bears repeating. Suppose, Sarfatti says, that time travel is possible. Then it is conceivable that a highly advanced civilization in the future could travel back in time and arrange the initial conditions of our universe in such a way that it would produce an environment that was hospitable to life. By doing this, it would guarantee its own existence. This sounds a bit like the Heinlein time-travel story in which a single individual was his and her own father, mother, and daughter. But of course it would certainly explain where our universe came from.

9

Why Do We Exist?

The Strange Properties of Water

In 1913, in his book *Fitness of the Environment,* the American physiologist Lawrence J. Henderson noted that if water did not possess certain strange properties, life could not exist on earth. For example, water is practically unique in that it expands, rather than contracts, when it freezes. If it acted in a more normal manner, Henderson pointed out, then the earth would freeze solid and life could not exist.

Because water expands on freezing, ice floats: A less dense substance will rise to the surface of more dense substances. If water contracted instead, as all "normal" substances do, then the coldest water in a lake or ocean would sink to the bottom and freeze. And of course any ice that formed on the surface would sink. As more ice formed during succeeding winters, the buildup would increase until finally all the bodies of water on earth would be frozen solid. Once this happened, the ice would never melt. After all, ice is a very reflective substance. If the oceans were covered with ice, much of the energy coming from the sun would be reflected back into space and the earth would become colder yet. Under such conditions, it would quickly grow far too cold to support life.

Indeed, water has so many anomalous properties that make it an ideal substance to provide a foundation for life that it is almost as though someone had consciously designed water with living organisms in mind. Another example is that water has a high specific heat. The amount of heat energy that is required to raise the temperature of a certain quantity of water by one degree is much greater than it is for most other substances. This means that, as water gains or loses heat, it will change temperature relatively slowly. As a result, the presence of water tends to stabilize the temperature of the environment. If the surface of the earth were not covered by large bodies of water, or if water did not have this property , temperature swings on the surface of the earth would be much greater, perhaps even too great for living organisms to endure.

In addition, water has the ability to dissolve an unusually great number of different substances. This characteristic makes it a superb medium for the various chemical reactions on which life depends. Water also has a high surface tension, which is important for the biochemical reactions that take place in living cells. And water has a high heat of vaporization; that is, a great deal of energy is required to turn a certain quantity of liquid water into vapor. This makes water an effective coolant, which is a property many living organisms make use of, as anyone who has ever worked up some perspiration or seen a dog with its tongue hanging out is aware.

THE SKEPTIC'S REPLY

A skeptic might reply that perhaps there is really nothing so remarkable about all this. If water didn't have certain special properties, then obviously creatures like human beings could never have evolved. But so what? Scientists don't know what kinds of life might or might not be possible. If water were not the kind of substance it is, then perhaps life forms based on sulfur compounds, or methane, or ammonia might evolve somewhere and they could construct similar arguments about their favorite life-giving substances.

Anyone who made such an objection might have a point. In fact, scientists tend not to make too much of the "magical" properties of water these days. But they do make similar arguments about other apparently accidental and serendipitous properties of the universe. For they are very much aware that a universe that can give rise to any kind of life is a very improbable thing.

THE IMPROBABILITY OF CARBON-BASED LIFE

In order to understand how miraculous it is that life did emerge in our universe, we can begin by showing that the existence of carbon-based life is very improbable. If our universe did not possess certain apparently accidental properties, there would be no such thing as carbon, or, at best it might exist only in very minute quantities. Elements such as oxygen and nitrogen, which are so common in our world and which are constituents of many important organic molecules, would not exist either.

The presence of carbon depends upon the existence of certain energy levels. According to the laws of quantum mechanics, an electron in an atom, or a particle in an atomic nucleus, cannot possess just any arbitrary quantity of energy. The energy of such a system is said to be quantized. This means that such a system can have only certain various definite amounts of energy, but that it cannot have anything in between. One can construct an analogy by imagining a bucket which could be filled a quarter full, or half full, or three-quarters full, or all the way to the top, but which could not hold any other quantities of water, such as one-eighth.

Because of this rule of atomic energy, when a particle leaps from one energy level to another, it is said to undergo a quantum jump (see Figure 11). The absorption or emission of energy has to be a jump because intermediate energy levels are simply not allowed. This, incidentally, is the reason why so many substances emit light of particular wavelengths. Each separate wavelength of light corresponds to a "jump" between two quantum levels.

One consequence of this is the fact that a system will tend not to absorb a packet of energy if that packet does not "fit." It's like

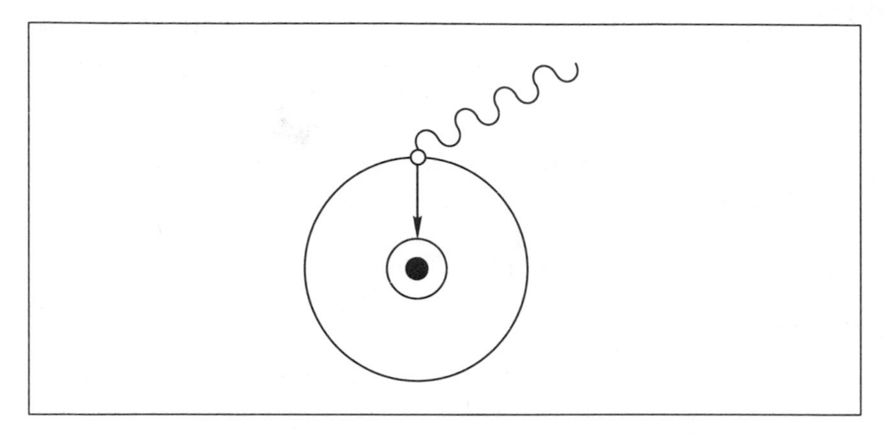

Figure 11 *In a quantum jump, an electron in an atom jumps from an outer orbit to one of lower energy, giving off a light wave in the process. Here, the light circle represents an electron and the dark circle the nucleus of the atom. The diagram should not be taken as a literal representation of subatomic reality. The uncertainty principle tells us that the electron's position cannot be pinpointed so exactly as this diagram seems to imply.*

that hypothetical bucket of water. Adding a quarter of a bucket would be easy. But anyone trying to add a sixth or a third of a bucket would have difficulties.

The subject of quantum levels, quantum jumps, and the absorption and emission of energy is one that could be discussed endlessly. But it is best if I do not attempt to go into the subject in too much detail. This would only succeed in obscuring what is really a very simple point: The chance that a nuclear reaction will take place depends upon the placement of the energy levels in the nuclei that are involved. A nuclear reaction is a reaction between previously existing nuclei or individual particles that causes a new kind of nucleus to be formed. An example would be the reaction which takes place in our sun which causes hydrogen nuclei to be bound together to make helium. The creation of a carbon nucleus from three helium nuclei is another.

If the energy levels happen to be just right, the reactions will take place very readily. If they are not, the reactions will take place only on very rare occasions. This is why some elements are so abundant and some are so rare: The energy levels of the nuclei make it easier to form some than others.

How does all of this relate to the subject of carbon-based life? Well, carbon is an element with a relatively simple structure. Its nucleus contains only six protons and six neutrons. The production of carbon is an intermediate step in the creation of oxygen and other heavier elements. If sufficient quantities of carbon were not created in our universe, these other substances would not be very abundant either.

But the existence of significant quantities of carbon seems to depend upon a fortuitous accident. Carbon and beryllium nuclei just "happen" to have energy levels in precisely the right place so that plenty of carbon is created in the nuclear reactions that take place in stars.

The energy levels in beryllium are important because beryllium, which has a nucleus with four protons and four neutrons, is one of the ingredients from which carbon is made. The process works something like this: Two helium nuclei collide. If they happen to strike one another with just the right energy, a beryllium nucleus is formed. Then, if this nucleus is struck by a third helium nucleus (again with the right energy), a carbon nucleus can be formed. And one must have carbon before oxygen is created. In fact, oxygen nuclei (which have a total of sixteen component particles) are generally created when nuclei of carbon (twelve particles) and helium (four particles) collide.

And how is it that carbon and beryllium have energy levels in just the places that they need to be in order to cause all the familiar elements to be synthesized? No one knows. It seems to be something that happened entirely by chance.

THE INFLATIONARY EXPANSION AGAIN

There are other characteristics of our universe that seem miraculous when we consider that life could not have appeared had these characteristics been the slightest bit different. For example, life would not be possible if the universe had not been expanding at precisely the right rate shortly after the big bang. If the expansion had been slower by a factor of even one part in a million when the universe was just a few seconds old, then, as we have seen in a previous chapter, the expansion would have been halted long before stars and galaxies formed, and the universe would have quickly collapsed in a big crunch. If, on the other hand, the expansion had been at one part in a million faster, the universe would have consisted of nothing but rapidly dispersing hydrogen and helium gas. The expansion of the universe would have caused the gas molecules to fly apart from one another at such a rate that they would never have had a chance to form the vast gas clouds that eventually condensed into stars and galaxies.

At one time, the fact that the expansion of the universe was so finely tuned seemed a great mystery. But today we know—or think we know—how this fine tuning took place. As we discovered in a previous chapter, the inflationary expansion produced a universe that was very nearly flat. And a flat universe is one that is right on the borderline between a too-rapid and a too-slow expansion. Furthermore, cosmologists and theoretical physicists generally tend to believe that the laws of physics support the conclusion that such an expansion was inevitable.

I must emphasize once again, however, that there is no empirical evidence to prove that an inflationary expansion took place in our universe. And of course this raises a whole series of questions: Would it be possible to have a universe in which no inflationary expansion took place? Do such universes exist? If so, should not one consider it an extraordinary coincidence that, in our universe, physical laws have the right character?

Or is it possible that there was no inflationary expansion, and that the expansion rate of our universe was another fortuitous accident?

DI-PROTONS

Another apparently fortuitous coincidence has to do with the strength of what is called the strong force, the force which binds such particles as neutrons and protons together in nuclei (as you recall, one of the four forces of nature). This strong force is just barely strong enough to bind a proton and a neutron together to make deuterium. But it is not powerful enough to create particles called di-protons, which would consist of two protons bound together. Such particles cannot form because the strong force cannot quite overcome the mutual repulsion of two positively charged protons.

But if the strong force were just a few percentage points stronger, then di-protons could be created, and the results would be catastrophic. If the creation of di-protons were possible, stars would not burn in the slow and steady manner that they do in our universe. In fact, there would be no such things as stars at all. Concentrations of primordial hydrogen and helium gas would produce nuclear reactions that progressed so rapidly that immense thermonuclear explosions would take place before stars would even form. And, since hydrogen would undergo reactions so readily, such a universe would be nearly 100 percent helium today.

One cannot be absolutely certain that life could not evolve in such a universe. After all, there might be life forms in this universe that are so unlike us that we are unable to imagine what they might be like. But, though one cannot entirely rule out the possibility of life in such a helium-filled universe, it certainly seems improbable.

More definite is the conclusion that no life could exist in a universe in which the masses of the neutron and proton were just slightly different, or in which the strong force was just a little bit weaker. If the strong force were 5 percent weaker than it is, deuterium would not exist, and there would be nothing that could bind a neutron and a proton together. Since the formation of deuterium is one of the steps by which hydrogen is made into helium, a universe with a strong force 5 percent weaker would not contain any stars. The hydrogen gas that it contained might condense under the influence of gravity in the same way that primordial hydrogen and helium did in ours. But the nuclear reactions that cause our stars to shine would never begin. At best, such a universe would be filled with brown dwarfs that had been heated by gravitational contraction. They might give off small amounts of heat radiation, but for the most part such a universe would be cold and dark.

The idea that forms of life that are entirely unlike us might evolve in such a universe seems unlikely. Whatever form life takes, after all, depends upon the flow of energy. A very high entropy universe that was filled with brown dwarfs does not look like a very suitable habitat for life of any kind.

A universe in which the ratio of the neutron and proton masses was reversed would be even worse. In our universe, the neutron is about 0.1 percent heavier than the proton. As a result, a neutron that is not bound in a nucleus will spontaneously decay into a proton and an electron. This does not happen because the neutron is a composite particle made up of a proton and an electron. It isn't. The proton and the electron are actually created in the decay process. One particle can become two particles because there is extra mass available.

If the proton were the heavier particle, the neutron would be perfectly stable, and protons would decay into neutrons and positrons when they were found in a free state. In such a universe, space would be filled with neutrons and little else. The positrons that were a product of the proton decay would undergo mutual annihilation with whatever electrons they encountered. Only neutrons would be left. It hardly seems conceivable that such a universe could have life. With only neutrons and nothing else, the complex structures on which life seems to depend could not be created.

THE BALANCE OF THE FORCES

As we have seen, the strength of the strong force must be just right to produce a universe that is suitable for the evolution of life. One can make similar statements about the gravitational, weak, and electromagnetic forces: The other forces of nature. For example, it is the gravitational force which caused primordial clouds of gas to condense into stars. If it were weaker than it is, stars might never have been created at all. A gravitational force that was too strong would produce results that were equally catastrophic, at least from our point of view: The universe might collapse in a big crunch too quickly for life to develop. Or, if it somehow managed to survive for a period of time comparable to that in which ours has existed, it might contain almost nothing but a lot of black holes.

Similarly, if the electromagnetic force, which binds molecules together, was too weak, solids and liquids could not exist, and the universe would contain nothing but gas. If, on the other hand, it was too strong, protons would repel one another too strongly, and complex atomic nuclei could not form.

The balance between the various different forces is important too. In fact, the last example that I gave is really an instance of this. The electrical repulsion between protons must not be so strong as to overcome the effects of the strong force. If it is, nuclei with more than one proton will not be created, and again, none of the reactions that sustain life could happen. But notice that it isn't the absolute strength of the repulsion that is so important; what matters is that it not be stronger than the attraction created by the strong force.

The ratio of the weak and gravitational forces is significant too. Certain nuclear reactions depend on the weak force, and also depend on the rate of expansion of the universe shortly after the big bang. The creation of such substances as helium and deuterium depends on conditions in the early universe. Altering the gravitational force would change the expansion rate, and thus the rate at which these reactions took place.

If the ratio of the weak and gravitational forces had been changed slightly, the universe would have emerged from the big bang either as 100 percent hydrogen or as 100 percent helium. There does not seem to be any special reason why a universe that was initially wholly hydrogen could not support life. On the other hand, it is hard to see how a universe that was 100 percent helium could.

THE DIMENSIONALITY OF SPACE

Yet another characteristic of our universe that is crucial to its ability to support life is the dimensionality of its space. There seems to be no particular reason why space should have three dimensions. It is quite easy to conceive of universes in which space is quite different. In fact, as we learned in the previous chapter, quantum cosmologists sometimes suggest that the dimensionality of space in our universe was one of the initial conditions, which suggests that it could well have had different characteristics.

We should, in fact, consider ourselves lucky that we live in a universe with three spatial dimensions. If the number were less or greater, life would almost certainly not exist. At least it is difficult to see how life could evolve in a universe with two spatial dimensions. For example, an animal could not possess a digestive tract that ran from one end of it to the other; it would be cut into two pieces. To be sure, attempts have been made to conceive of what life might be like in a two-dimensional world. However, these have something of the character of mathematical games. As far as I know, no one has seriously suggested that two-dimensional life would be possible.

Similarly, a universe that had more than three spatial dimensions would also almost certainly be inimical to life. In such a universe, many of the laws of physics would have a different character than they do in ours (this is also true of a universe with two spatial dimensions). For example, if the dimensionality of space were four or greater, then stable planetary orbits would not be possible. If something resembling a planet did manage to form, it would follow a path that caused it to spiral into its sun.

THE ANTHROPIC PRINCIPLE

The fact that the universe should be so hospitable to life certainly seem a perplexing puzzle, dependent as it is on so many apparently accidental factors. Attempting to solve this puzzle has naturally led to a great deal of speculation, much of which is more philosophical than scientific in character. Nevertheless, scientists rather than philosophers have carried on most of the debate about such questions.

Perhaps there is nothing very surprising about this, since these questions have arisen in the context of cosmological theory rather than that of philosophical debate. So, for better or for worse, some contemporary scientists have been forced to become philosophers. In an attempt to deal with such questions, these scientists have developed what is known as the anthropic principle, which can be expressed both in a "weak" and a "strong" form.

The weak anthropic principle has been stated by the British physicist Brandon Carter as follows: "What we can expect to observe must be restricted by the conditions necessary for our presence as observers." In other words, if the universe did not have certain characteristic properties, we would not be here to see it.

At first, the weak principle sounds almost like a platitude. But when one begins to consider its implications more deeply, it takes on more significance. The use of the weak principle does answer certain questions.

For example, how does it happen that we live in a universe that is about 15 billion years old? One could answer by saying that in a universe that was too young or too old, there would not be any conscious observers. If there is to be intelligent life, it appears that the universe has to be a few billion years old, at least, since it takes time for galaxies and stars, and then life, to evolve. And it is also unlikely that there will be any conscious observers when the universe attains an ago of a trillion years. At that time, it will be cold and dark and all but a few dim stars will have died out.

There seems to be nothing very remarkable about the weak anthropic principle. Nevertheless, the idea has had its critics. And, significantly, the principle has been criticized on philosophical rather than scientific grounds. For example, at a symposium which was held at the 1988 meeting of the American Association for the Advancement of Science, the weak anthropic principle was branded as a kind of "cosmic narcissism" by physicist Heinz Pagels. In criticizing it, Pagels drew on certain ideas that had been proposed by the Austrian-British philosopher Karl Popper. He made no attempt to disprove it in the manner that one would try to disprove a scientific theory.

In his 1931 book, *The Logic of Scientific Discovery,* Popper had argued that every scientific hypothesis must be falsifiable. If it is, then it cannot be called "scientific." For example, the statement "God exists" may or may not be true. But true or not, it is not a scientific hypothesis because it is not susceptible to disproof. Einstein's general theory of relativity, on the other hand, is scientific because it makes predictions that can be tested by experiment.

According to Pagels, the weak anthropic principle is not scientific because it cannot be disproved under any conceivable circumstances. I suspect that it is necessary to agree with him on this point. However, even though the weak principle is not really a scientific idea, it does seem capable of answering certain questions that arise within scientific contexts. Not only does it tell us why the universe we see should have a certain approximate age, but it may also tell us why we should find ourselves living on a water-covered planet that revolves around a smallish, long-lived star, and so on. This may be the only kind of place where life exists. If indeed there is nothing very "scientific" about the weak principle, then perhaps that is only an indication that there are points where the border between science and philosophy becomes somewhat blurred. And perhaps that is not such a bad thing.

THE STRONG ANTHROPIC PRINCIPLE

If the weak anthropic principle seems somewhat philosophical, the strong version is positively metaphysical. The strong anthropic principle is an alternative way of interpreting the same phenomenon (the existence of conscious life in the universe), and it has been stated by Carter as follows: "The universe must be such as to admit the creation of observers within it at some stage." In other words, the universe has to be of such a character as to provide an environment in which conscious beings can evolve.

To someone steeped in Western scientific tradition, such an idea has metaphysical or theological overtones which we do not associate with science. It is the kind of idea that was sometimes discussed during the eighteenth and nineteenth centuries, when it was fashionable to see the hand of a Creator in natural phenomena.

I am not claiming that the strong anthropic principle necessarily has religious implications. Nevertheless, it does seem reminiscent of the so-called argument from design. This argument, which was once quite popular and which is still sometimes cited in religious literature, is based on the idea that the existence of God is revealed by the wonders of the natural world. The world, and the life that populated it, could not have come about by accident. Thus its existence could be viewed as a proof of the existence of a deity.

Today, the argument is no longer used by theologians, and many scholars consider that it was demolished by the eighteenth-century German philosopher Immanuel Kant. Nevertheless, one does have to admit that one way of explaining why the universe should be so hospitable to life would be to answer that God designed it that way.

This is not an answer that would appeal to many scientists, however, whether they happened to be religious or not. As a group, scientists are just like everyone else in that some believe one religious creed or another, whereas others are atheists or agnostics. However, even those who believe in a Creator are reluctant to ascribe characteristics of the universe to His conscious design. Even the most devout tend to feel that questions that arise in a scientific context should be answered in a scientific context, and that theological considerations should be avoided.

INTERPRETING THE STRONG ANTHROPIC PRINCIPLE

In their book *The Anthropic Cosmological Principle,* scientists John D. Barrow and Frank J. Tipler suggest that, if the strong anthropic principle is true, then there are three different ways of interpreting it.

The first possible interpretation is that the universe was deliberately designed to be hospitable to life. According to Barrow and Tipler, this view is not open either to proof or disproof because it is religious in nature. However, I am not so sure that it has to be. There exist some (admittedly very unlikely) explanations for this interpretation that have nothing to do with religion. For example, scientists of an advanced technological civilization might be able to create universes, and to set the initial conditions so that life is likely to evolve in them. After all, during the period before an inflationary expansion begins, a newly born universe is likely to be nothing but a tiny ball of spacetime that is empty, or nearly empty, of matter. For all we know, our universe may be some graduate student's experiment gone awry.

The second possibility that Barrow and Tipler cite is that observers are necessary to bring the universe into being. In other words, a universe cannot come into existence if it is incapable of producing conscious beings that can observe it. Barrow and Tipler attempt to give this somewhat mystical-sounding view a scientific foundation by relating it to certain recently developed interpretations of quantum mechanics. Here I will have to refer the interested reader to their book. The question of the proper interpretation of quantum mechanics is quite a complex subject, one on which many books have been written. Furthermore, the topic is more philosophical than scientific in nature (when one asks what numerical measurements quantum mechanics predicts, that is science; when one asks what quantum mechanics means, that is philosophy), and I don't have the space to go into such matters here. I will only comment that I personally find speculation about observer-created universes to be metaphysical in the extreme, and I am not sure that relating it to ideas in quantum mechanics makes it any less so.

INFINITE WORLDS

To my mind, the third possibility cited by Barrow and Tipler is the most straightforward and logical one. I believe that one can interpret the principle to mean that there are a very large number, perhaps an infinite number, of different universes. In such a cases there would be nothing surprising about the fact that conscious life exists in our own. After all, there would also be countless other universes where, because conditions were not precisely right, life never evolved. In some of these universes, there might never have been an inflationary expansion. Others may never have stopped expanding at an inflationary rate. In some of these universes, stars and galaxies may never have formed. And in others, there might be no elements heavier than hydrogen and helium.

The point is that, if there are really a large (or infinite) number of different universes, all the problems associated with the improbability of a universe hospitable to life disappear. In such a case, our universe would still be a very improbable one. But there would be nothing striking about that because all the other, more likely, kinds of universes would exist too. But of course there would be no one there to see them.

It should be noted, however, that this kind of idea only seems to work if we make the assumption that the laws of nature can vary from one universe to another. For if there were a large number of different universes, and they all had physical laws similar to those which operate in ours, then they would most likely all contain life, and the question of why a universe should be so hospitable to life would remain.

All of this surely sounds like wild speculation. However, it is speculation that can be placed in a real scientific context. After all, scientists are no longer content with discovering what the laws of physics are. They are now also asking why physical laws have the form they do.

There are only two possible answers: One is that there may exist numerous universes and that the laws of physics may not be the same in all of them. The other is that there may be only one set of laws that is logically consistent. They may find that there is some reason why there have to be just four basic forces. There may be some reason why a certain set of subatomic particles exists, whereas others that are theoretically possible don't. They may discover that there is a reason why particles have certain specific masses. It may be that, in any universe, for some logical reason, a neutron *must* have a mass slightly larger than that of a proton, for example. And if this turns out to be the case, the third interpretation of the strong anthropic principle will have been disproved because logic could allow for only one type of universe.

From time to time in this book, I have mentioned superstring theories. Since this is a book on cosmology, not particle physics, I have not attempted to discuss such theories in detail. I should point out, however, that what the superstring theorists are seeking is a "theory of everything," a theory from which all the known laws of physics could be derived. If such a theory is eventually found, it is possible that scientists will discover that the reason that the four fundamental forces have certain characteristic strengths, that certain particles have certain masses, and so on, is that nature had no choice.

On the other hand, it may be discovered that there are many elements in the physical laws that are completely arbitrary. Scientists may find that the strengths of the forces, the masses of the elementary particles, and perhaps even the dimensionality of space may depend entirely fortuitously upon the initial conditions of the universe.

It now appears that whereas the somewhat modest-sounding weak anthropic principle is basically philosophical in nature, the more metaphysical-sounding strong principle can be tested by empirical observation. If a successful superstring theory is ever developed or, alternatively, if the quantum cosmologists ever attain their goals, then we will most likely be able to say whether the third interpretation of the strong principle is true or false.

10

Knowing the Mind of God

When the results of the COBE measurements of the cosmic microwave background were announced, some of the scientists who had worked on the project put their findings in a religious context. "If you're religious, it's like seeing God," said George Smoot of the temperature fluctuations his team had measured. Team member John Mather spoke of seeing parallels between the satellite's measurements and the biblical story of creation.

These comments generated an immediate response from other members of the scientific community, who objected that comments of the type were misleading. Smoot's and Mather's critics said that although the desire to convey a sense of awe about the new findings was understandable, references to God were a mistake, and possibly even an abuse of scientific authority. Scientists should not make such remarks, which would only cause the public to confuse scientific with religious ideas, or to incorrectly infer that scientific and religious views of the creation were somehow related.

The critics may have had a point. However, while voicing such criticisms they were guilty of feigning ignorance about the fact that many scientists speak about God these days, even when they are totally irreligious. For example, Stephen Hawking is an outspoken atheist. And yet he speaks unabashedly of knowing the "mind of God" in his book *A Brief History of Time.* In the conclusion to his book, speaking of what is likely to ensue if physicists succeed in discovering a "theory of everything," he writes: "Then we shall all, philosophers, scientists and just ordinary people, be able to take part in the discussion of the question of why it is that we and the universe exist. If we find the answer to that, it would be the ultimate triumph of human reason—for then we would know the mind of God." Similar remarks have been made by the American physicist Freeman Dyson. Ironically, Dyson is one of Hawking's philosophical opponents in that he does not believe that a complete theory, which would explain all the other laws of physics, will ever be found. And yet Dyson's opposition to Hawking has not prevented him from making his own references to God. He writes: "If it should turn out that the whole of physical reality can be described by a finite set of equations, I would be very disappointed," Dyson has said. "I would feel that the Creator had been uncharacteristically lacking in imagination."

EINSTEIN'S GOD

Speaking of God in a scientific context certainly did not begin with Smoot, with Hawking, or with Dyson. In the time of Isaac Newton, for example, scientists habitually spoke of God's designs. And in the modern era, Albert Einstein was notorious for his tendency to speak of God's intentions. Though Einstein had lost his faith in the personal God of Judaism and Christianity while he was still an adolescent, he often spoke of "God" or "the Lord" when he wanted to make statements about the natural order. The most famous of these statements was, of course, "God does not throw dice," which was an expression of his dislike of the randomness of quantum mechanics. "The Lord is subtle, but not malicious," he would remark when he wanted to make the point that, although natural laws might sometimes be difficult to decipher, the universe was basically understandable. On yet other occasions, he spoke in an almost mystical manner, commenting that his God was one who revealed Himself "in the harmony of all that exists."

On some occasions, Einstein might have spoken of God's intentions with a little too much confidence. Once, when Einstein had apparently repeated his assertion that "God does not throw dice" one too many times, the Danish physicist Niels Bohr somewhat testily replied, "Stop telling God what to do." However, Bohr, who argued with Einstein for years about the correct interpretation of quantum mechanics and who was a severe critic of some of Einstein's ideas, understood quite well that his colleague spoke of God in order to communicate his sense of awe when contemplating the order of the universe.

If scientists, even irreligious ones, speak of God with increasing frequency today, it may be because—like Einstein—they know of no other way to adequately express the awe they feel when contemplating some of the things that science has discovered, or is on the verge of discovering. Today, the most advanced scientific fields, such as particle physics and cosmology, appear primed to tackle questions that were once considered to be purely in the province of metaphysical philosophy or religion.

Although few definitive answers have yet been found, scientists have begun to find ways to deal with such questions as "Where did the universe come from?" and "How did time begin?" They have also begun to ask, "Could the universe be significantly different from the way it is?" And though they don't yet know the answer to this query, they do know what lines of research must be pursued if the problem is to be solved. If quantum cosmology successfully relates the initial conditions of a universe to its structure at later times, then we will at least know what various kinds of universes are possible.

When George Smoot and his colleagues found temperature fluctuations in the big bang, it was like seeing God in a sense, or perhaps like seeing God's fingerprints. After all, these "bumps in the big bang" presumably had their origin in microscopic quantum fluctuations, and were later enlarged into visibility by the inflationary expansion. Being able to see a relic of this primordial quantum world was like peering into the foundation of all reality.

It must be emphasized, however, that scientists are not seeing God's hand in cosmic data in any literal sense. The latest discoveries have not brought science and religion closer together, and they are not likely to do so. No matter how deeply science probes into the nature of physical reality, it will continue to explain the things that it discovers in scientific terms. Those who are so inclined can continue to imagine the presence of a Creator behind the scenes if they so desire. But God is not something that is encountered in the physical data.

COSMIC QUESTIONS

I called this book *Cosmic Questions* because I wanted to emphasize that there is a whole series of metaphysical-sounding questions that scientists now think they will eventually be able to answer. In some cases, they are probably very close to solutions. Although relatively few definitive answers have been found so far, there can be no doubt that dramatic new discoveries lie ahead. As I have pointed out on more than one occasion, we are currently witnessing the beginning of a golden age of cosmology. I, for one, will be very surprised if our conception of the universe a decade from now is not very different from what it is at present.

What are these cosmic questions? Some of them are cited in the chapter headings. Others have been discussed in the text. And there are still others that have not even been formulated as of yet. As new discoveries are made, new questions will inevitably arise. Here is a list of some of the questions that I personally find most intriguing:

What is the universe made of?

How did it begin?

How will it end?

What is time?

Would a universe in which time had a different character than it does in ours be possible?

Are the laws of nature that we observe the only ones that are logically possible, or do they depend upon the initial conditions of the universe?

The same question phrased differently: Did God have any choice when He devised the physical laws by which our universe operates?

Are there other universes?

Could the laws of nature in these universes be different from what they are in ours?

Why is our universe so hospitable to life?

Will a universe like ours come into existence again and again and again? If so, will life forms that resemble us evolve countless times in the future?

Glossary

absolute zero −273°C: the lowest possible temperature, and the temperature at which all molecular motion ceases.

anthropic principle The principle which attempts to deduce certain facts about the universe from the fact that we exist and can perceive it. the anthropic principle has a "weak" and a "strong" form.

antimatter *See* **antiparticle**.

antiparticle For every particle, there exists an antiparticle. When a particle and an antiparticle encounter one another, they undergo mutual annihilation and are transformed into energy. Antimatter, which has not been observed to exist in nature, would be a kind of matter made up of antiparticles. Its properties would be similar to those of matter.

arrows of time The fundamental laws of physics do not distinguish between past and future. Yet nothing could be more obvious than the fact that the two directions of time are quite different. The "arrows" of time are phenomena that distinguish between the two time directions. They include the direction of increasing entropy, the expansion of the universe, the "arrow" of electrodynamics, the decay of the kaon, and human consciousness. It is not known how the various different arrows of time are related. *See also* **time reversibility**.

baryons Heavy particles such as protons and neutrons. Other kinds of baryons exist, but they are observed only in the laboratory.

baryonic matter Matter that is made up of baryons (protons and neutrons): the "ordinary matter" of our everyday world. *See* **baryons**.

big bang The initial, explosive expansion from a very hot, compressed state that is thought to have created our universe. The phrase was invented by a proponent of a competing (and now discredited) theory, who used it as a term of derision.

big crunch It is not known whether the expansion of the universe will ever slow down and reverse itself. If it does, a state of contraction will set in, and the universe will eventually destroy itself in a big crunch, an analogue of the big bang.

black hole The collapsed remnant of a dead star in which gravity is so strong that nothing, not even light, can escape from it.

blueshift *See* **redshift**.

brown dwarf When gravity causes a ball of primordial hydrogen and helium gas to contract, the resulting pressure causes the gas to heat up. If the temperature grows high enough, nuclear reactions will begin in the core of the ball of gas, and it will become a star. But if the ball of contracting gas has a mass that is less than 7 or 8 percent of the mass of our sun, this will never happen, and a kind of "failed star" called a brown dwarf will be the result. Brown dwarfs radiate some heat, but since there are no nuclear reactions to produce energy, they give off little or no light. Some of the dark matter in the universe may consist of brown dwarfs. *See also* **dark matter**.

COBE NASA's Cosmic Background Explorer satellite. In 1992, instruments placed on the COBE satellite detected "bumps" or temperature fluctuations in the big bang. Astronomers had been searching for such fluctuations for years.

closed universe A finite universe in which space closes in upon itself. Though finite, it has no boundaries. In such a universe, the expansion must eventually halt and be followed by a phase of contraction. *See also* **flat universe** and **open universe**.

cold dark matter *See* **dark matter**.

conservation of energy According to this principle, which was developed during the nineteenth century, energy can be neither created nor destroyed: it can only be changed into other forms of energy. Einstein's famous equation $E = mc^2$ forced scientists to modify this law, since it showed that matter and energy can be transformed into one another.

cosmic microwave background radiation A background of microwaves that constantly falls on the earth from every direction of space. It is a remnant of the radiation that was emitted in the big bang.

cosmic rays Charged particles (not radiation) that impinge on the earth's atmosphere from all directions in space. They consist primarily of protons but also include various kinds of heavier atomic nuclei.

cosmic string One of several different kinds of structural flaw in spacetime that might have been created around the time of the inflationary expansion, roughly analogous to the flaws in a crystal such as a diamond, or cracks in a sheet of ice. If cosmic strings exist, they would be very long, very thin, and very massive. Loops of cosmic string could have been the gravitational "seeds" around which galaxies formed.

cosmological constant A constant that Einstein introduced into his equations of gravitation. It corresponds to a cosmic repulsive that would pervade the universe. The cosmological constant should theoretically be very large. In reality, it is either zero or so close to zero that it cannot be measured. This discrepancy between theory and observation is not yet understood.

critical density The figure for the average mass density of the universe that determines whether the universe is closed or open. If the density is greater than this amount, the universe is closed and its expansion will eventually halt. If it is less than this amount, the retarding effects of gravity will not be sufficient to slow the expansion to this extent, and the universe will continue to expand forever. *See also* **closed universe** and **open universe**.

curved space This term can be a bit misleading. Space, after all, is not an object that can be bent in the manner that, say a spoon, can be bent. When scientists speak of "curved space" they mean that the geometry of space has changed. For example, in flat, or Euclidian, space the angles of a triangle always add up to 180 degrees. In curved space, the sum of the angles of a triangle may be either greater or less than 180 degrees.

dark matter The nonluminous matter that cannot be observed but that constitutes at least 90 percent of the universe. Its composition is unknown. At least some of it may consist of light particles such as the neutrino. This form is referred to as hot dark matter because such particles would have emerged from the big bang at high velocity. Relatively heavy, slow-moving particles would be called cold dark matter. Dark matter might also exist in a number of other forms, discussed in the text.

deuterium A form of hydrogen with a nucleus composed of a proton and a neutron rather than a single proton. The term *deuterium* is also somewhat loosely applied to the deuterium nucleus itself.

di-proton A theoretical particle composed of two protons. Such particles do not exist because the electrical repulsion between the two protons is too strong. However, they could exist in a universe in which the repulsion was slightly weaker or the force binding the protons together was stronger.

electrodynamics The field of physics that deals with the behavior of moving electrical charges and with such phenomena as electromagnetic radiation.

energy level According to quantum mechanics, electrons in atoms can possess only certain specific quantities of energy: they cannot have anything in between. Discrete energy levels are also a characteristic of atomic nuclei. *See also* **quantum jump**.

entropy Often equated with the "disorder" of a system, but can also be defined in terms of disequilibrium. A low-entropy system would then have a large amount of disequilibrium, or energy differences between different parts of the system. In a high-entropy system, the different parts have come into equilibrium with one another. Since these different parts possess roughly equivalent levels of energy, little or no useful work can be done; energy can no longer flow from one place to another. *See also* **second law of thermodynamics**.

false vacuum According to quantum mechanics, even empty space can possess energy. When space is not in its lowest energy state, it is said to be in a false vacuum. The entire universe may originally have been in such a false vacuum state, and the resulting transition to a lower energy state may have been what drove the inflationary expansion. *See also* **inflationary expansion** and **true vacuum**.

flat universe One in which the average curvature of space would be zero. Such a universe would lie exactly on the borderline between an open and a closed universe. It is thought that our universe is very nearly flat. *See also* **closed universe** and **open universe**.

forces The four basic forces in our universe: the gravitational, the electromagnetic, and the strong and weak nuclear forces. The weak and electromagnetic forces can be described as two different aspects of a single "electroweak" force. Scientists would like to find a theory that would explain all four forces within a single theoretical framework.

galactic halos The halos of dark matter that surround galaxies. Though this matter cannot be seen, its gravitational effects can be observed. *See also* **dark matter**.

general theory of relativity Einstein's theory of gravitation, propounded in 1915. It is believed that this theory remains valid except under very extreme conditions, and it is used to describe the evolution of the universe from a time very near the beginning (10^{-43} seconds after the creation of the universe) to the present day. The theory does have its limits, however, and scientists hope to eventually find a way to replace it with a theory of quantum gravity.

globular cluster A compact collection of stars outside a galaxy. The stars in these clusters are generally very old. Our galaxy is surrounded by about 200 such clusters of about a million stars each.

grand unified theories (GUTs) Theories that attempt to combine descriptions of the electromagnetic and strong and weak forces within a single theoretical framework. These theories are speculative, and no one knows which, if any, of them is most likely to be true.

gravitational lens According to Einstein's general theory of relativity, a massive object such as a galaxy can bend light to produce multiple images of a distant object, such as a quasar, that is behind the galaxy. This effect has been observed by astronomers. Dark matter is also observed to create gravitational lens effects.

hot dark matter *See* **dark matter**.

Hubble constant A number which measures the rate at which our universe is expanding. Because of uncertainties inherent in measurements of the distance of galaxies, the Hubble constant is uncertain by a factor of about 2. Some astronomers favor a Hubble constant of about 50 (corresponding to an age of about 15 to 20 billion years for our universe), whereas others think that the constant is closer to 100 (implying an age of 7 to 10 billion years). The controversy has not yet been resolved.

imaginary time According to a theory developed by physicists Stephen Hawking and James B. Hartle, called the no-boundary proposal, the universe may have begun in imaginary time, *imaginary* being used in a technical, mathematical sense. According to this theory, time originally had a spatial character and the universe began as a sphere of four spacelike dimensions. Thus originally there was no time in the sense in which we ordinarily understand the term. If the no-boundary proposal is correct, the universe did not have a beginning in time.

inflationary expansion A state of extremely rapid expansion supposedly undergone by the early universe. If this theory is correct, our universe is very nearly flat and its average mass density is very close to critical density. *See also* **critical density** and **flat universe**.

kaon A subatomic particle, formerly called the K meson, which provides an exception to the principle that all of the fundamental laws of physics are time-reversible. If a videotape could be made of kaons as they decayed into other particles and the tape was played later, it would be possible to tell whether the tape was being run in the forward or backward direction. This is not true of any other particle decay. It is not known why the kaon exhibits this characteristic; it is not a constituent of ordinary matter. *See also* **arrows of time** and **time reversibility**.

local group A small cluster of about twenty galaxies that includes our own galaxy and the great galaxy in Andromeda.

magnetic monopole A hypothetical particle 10^{16} times heavier than the proton which would have the appearance of an isolated north or south magnetic pole. If magnetic monopoles exist, they are unlike all other particles. In a certain sense they wouldn't be particles at all, but rather pointlike flaws in spacetime.

megaparsec The unit generally used by astronomers to measure distances: approximately equal to 3 million light years.

negative pressure The quality exhibited by an innate tendency to contract. Negative pressure may, somewhat paradoxically, have played a role in the inflationary expansion of the universe. *See also* **inflationary expansion**.

neutrino A very light uncharged particle whose mass is either zero or too small to be measured. There are three different varieties of neutrino.

no-boundary proposal *See* **imaginary time**.

open universe A universe in which the curvature of space is such that the universe does not close upon itself. Thus an open universe is infinite in extent. It also differs from a closed universe in that the expansion never slows down to zero. *See also* **closed universe** and **flat universe**.

phase transition A change of matter from one state to another, such as the melting of a block of ice or the boiling of water. According to quantum mechanics, spacetime can also undergo phase transitions as it changes from one energy state to another. Phase transitions might have played an important role in the evolution of the universe.

positron The antiparticle of the electron, having the same mass as the electron but possessing a positive, rather than negative, electrical charge. When a positron and an electron encounter one another, mutual annihilation is the result. The mass of the two particles is converted into energy, and a pair of gamma rays appears in their place.

primordial black hole Primordial black holes, sometimes called "mini black holes" are theoretical objects that might have been formed early in the history of the universe. They would be much smaller than the stellar black holes that are remnants of dead stars. If they existed they would be difficult to detect. Primordial black holes could be a component of dark matter. *See* **dark matter.**

quantum cosmology The field which seeks to find theories to explain the relationships between the initial conditions of the universe and the structure of the universe today.

quantum fluctuation According to quantum mechanics, the tendency of particle-antiparticle pairs (an electron and a positron, for example) to "pop" into existence out of nothing and then quickly disappear again. Some scientists believe that our universe may have begun when a tiny ball of spacetime came into existence through a similar process. *See also* **virtual particle**.

quantum jump A sudden transition from one energy state to another.

quantum mechanics The theory used to describe the behavior of subatomic particles. It is one of the most successful theories science has ever known and provides a foundation for all of modern physics.

quark The theoretical constituents of the proton, the neutron and other heavy particles (not of particles such as the electron and the neutrino).

quark nuggets It has been suggested that some dark matter might consist of conglomerations of primordial quarks. If quark nuggets exist, they might range in size from a diameter of about a twenty-fifth of an inch to one meter (about a yard). However, quark nuggets have never been observed.

quasar The luminous core of a young galaxy. It is believed that the brightness of quasars can be attributed to the radiation emitted by matter that falls into supermassive black holes in the quasars' centers. *See also* **black hole**.

red giant As the nuclear fuel of an aging star begins to become exhausted, it burns its remaining fuel at an ever-increasing rate and expands into a hot, bright, red giant star. Our sun will become a red giant in about 5 billion years. Its diameter will be

400 times greater than it is now, and its light output will have increased by a factor of 10,000. When a red giant's fuel is completely exhausted, it may become a neutron star, collapse into a black hole, or shrink into a small, dim white dwarf. Our sun will become a white dwarf. *See also* **black hole** and **white dwarf**.

redshift When an object moves away from an observer or the observer is moving towards it, the wavelengths of the light that it emits are lengthened. Since the wavelengths of red light are longer than those of any other part of the visual spectrum, there will be a shift toward the red.

second law of thermodynamics The law stating that the entropy of any isolated system can only increase. As a system's entropy becomes greater, its ability to perform useful work diminishes. The universe is an isolated system. Consequently its ability to support life will inevitably diminish over a time frame of tens of billions or hundreds of billions of years. *See* **entropy.**

shadow matter A hypothetical form of matter that would interact with ordinary matter only through mutual gravitational interaction. If shadow matter did exist, it could be neither seen nor felt. But it would have gravitational effects, and it could contribute to the mass density of the universe.

singularity The point of infinite density that would be created if a quantity of matter were compressed by gravity into a mathematical point. Though the general theory of relativity predicts the existence of singularities (in the center of a black hole, for example), it is not likely that they exist, because quantum effects would probably ensure that the matter density never became infinite.

solar neutrino problem It is possible to calculate how many neutrinos should be produced in the nuclear reactions that fuel our sun and to estimate how many of those neutrinos should be falling on the earth. However, when attempts are made to detect these neutrinos experimentally, scientists find fewer than expected. This discrepancy between theory and experiment, has not yet been explained. *See also* **neutrino**.

spacetime A word physicists use to describe the three dimensions of space and the one dimension of time. The acceptance of Einstein's theories caused the word to gain currency because, in Einsteinian physics, space and time interact in a way that they do not in Newtonian mechanics. Nevertheless, it would be perfectly correct to speak of spacetime in the context of Newton's theories also.

special theory of relativity Einstein's theory, propounded in 1905, that deals with the behavior of bodies which travel at velocities approaching that of light. It is this theory which produced the famous equation $E = mc^2$. It is also special relativity which implies that, if something could travel faster than light, it could also move backwards in time. Most scientists believe that such faster-than-light travel is impossible.

supernova The explosion created when a large star, one with more than about three or four times the mass of our sun at the end of its life, undergoes gravitational collapse when its nuclear fuel is exhausted and is transformed into a gigantic nuclear bomb. So much energy is released in such an explosion that, for a brief time, the dying star may be billions of times as bright as our sun. *See also* **type Ia supernova**.

superstring theories Many theoretical physicists suspect that all known particles might consist of vibrating loops known as superstrings. If they exist, superstrings would be many orders of magnitude smaller than a particle such as an electron or a proton, and it would be impossible to observe them anytime in the foreseeable future. However, advocates of superstring theory point out that, if a successful theory was found (none has been yet), it might provide a unified explanation of all four forces of nature. *See also* **forces** and **theory of everything**.

textures Hypothetical defects in spacetime that might have been created during phase transitions. If textures exist, they would contain twists analogous to those in a rubber band, and would thus have more complicated forms than such objects as cosmic strings. *See also* **cosmic string** and **phase transition**.

theory of everything A theory from which all of the other laws of physics could be derived. It is not known whether it is possible to find such a theory. However, many scientists hope that a superstring theory will eventually prove capable of providing such all-embracing explanations.

time reversibility All of the basic laws of physics, including those which govern gravity, the emission and absorption of radiation, and all of the common atomic and nuclear reactions, are time-reversible: They do not distinguish between past and future. This means that, ordinarily, any fundamental process can also take place in reverse without violating any physical law.

true vacuum The lowest energy state of "empty" space. *See also* **false vacuum** and **phase transition**.

type Ia supernova The type of supernova that occurs when gravity causes material to be drawn from the surface of a red giant star onto a white dwarf companion. When enough mass has accreted on the surface of the white dwarf, the dwarf will be suddenly transformed into a massive nuclear bomb. Theoretically, all type Ia supernovas should occur at exactly the same mass, so they should all be equally bright. Obviously, they can only take place in a binary star system.

uncertainty principle The principle, discovered by the German physicist Werner Heisenberg, which states that the momentum and position of a subatomic particle cannot be simultaneously determined, or even defined. The uncertainty principle implies that there should be a similar relationship between energy and time. This implies in turn that, for very short periods, there is enough energy available to create particles that could otherwise not exist. *See also* **virtual particle**.

virtual particle A particle-antiparticle pair that "pops" into existence even when the energy that would normally be required to create it is not available. The energy "debt" thus created must be quickly paid back, and these virtual particles disappear almost as soon as they are created. Virtual particles are not a theoretical chimera. They have real physical effects which have been experimentally measured to great accuracy. *See also* **quantum fluctuation**.

weakly interacting massive particle (WIMP) A term which embraces a large number of subatomic particles which have been proposed on theoretical grounds but which have never been observed. Cold dark matter would presumably be made up of WIMPs. *See also* **dark matter**.

white dwarf A dead star in which nuclear reactions no longer take place. White dwarfs are observed by astronomers because their residual heat causes them to continue to glow.

wormhole A hypothetical pathway that would join two widely separated regions of space. If other universes exist, it is conceivable that our universe could be connected to them by wormholes. Wormholes have never been observed, and their dimensions may be so small (about 10^{-33} centimeters) that scientists may never be able to see them. Nevertheless, like virtual particles, their existence could have observable effects.

X particle According to the grand unified theories, a hypothetical particle, which would decay into particles more often than it would decay into antiparticles. If the X particle exists, it would explain why there is so much more matter than antimatter in the universe. The X particle has not been detected. It is thought to be so massive, 10^{14} or 10^{15} times heavier than a proton, that it could not be produced in any existing particle accelerator.

Index

Numbers in italics refer to figures and tables.